Carbon Inventory Methods: Handbook for Greenhouse Gas Inventory, Carbon Mitigation and Roundwood Production Projects

ADVANCES IN GLOBAL CHANGE RESEARCH

VOLUME 29

The titles published in this series are listed at the end of this volume.

N. H. Ravindranath • Madelene Ostwald

Carbon Inventory Methods

Handbook for Greenhouse Gas Inventory, Carbon Mitigation and Roundwood Production Projects

Springer

N. H. Ravindranath
Centre for Sustainable Technologies &
Centre for Ecological Sciences
Indian Institute of Science
Bangalore
India

Madelene Ostwald
Centre for Climate Sciences and Policy
Research Linköping University &
Earth Sciences Centre
Göteborg University
Sweden

ISBN-13: 978-1-4020-6546-0 e-ISBN-13: 978-1-4020-6547-7

Library of Congress Control Number: 2007936020

Printed on acid-free paper.

9 8 7 6 5 4 3 2 1

springer.com

Preface

Global awareness of environmental issues has increased on an unprecedented scale. Deforestation, land degradation, desertification, loss of biodiversity, global warming and climate change are some of the environmental issues linked directly to terrestrial ecosystems, both natural and human-managed. Forests, grasslands and croplands constitute over 63% of the global land area. Terrestrial ecosystems play a critical role in the global carbon cycle. Global rise in demand for food, fodder, fuel and roundwood is increasing the pressure on land-use systems, and conservation and sustainable development of land-use systems are critical for meeting those demands sustainably and stabilizing CO_2 concentration in the atmosphere to mitigate global climate change.

My interest in carbon flows in forest ecosystems was initiated about 15 years ago with the global concern about the contribution of the growing CO_2 emissions to climate change. My first paper about carbon flows was published in 1996 in *Climatic Change*; since then my interest and the global interest have only increased (R).

My participation in the preparation of IPCC reports on *Greenhouse Gas (GHG) Inventory Guidelines* as a member of the team of authors responsible for the *Revised 1996 IPCC Guidelines* (Land Use Change and Forestry – LUCF), 2003 (Land Use Land-Use Change and Forestry – LULUCF) and 2006 (Agriculture, Forests and Other Land Uses – AFOLU) cemented my interest in carbon inventories, which was stimulated further by my contribution to the *Special Report of IPCC on Land Use Land-Use Change and Forestry* (LULUCF) and the third and fourth assessment reports of IPCC. Participation in preparation of the IPCC GHG inventory guidelines, their application for carbon inventory and review of their use across a large number of countries sowed the seeds of a carbon inventory methods manual, since I knew what IPCC guidelines provided and what was needed by the inventory experts to prepare the inventory for land-use systems, as all countries are required to prepare a carbon inventory for land-use categories as a part of their GHG inventory (R).

Secondly, in the global efforts to address climate change, there is huge interest in the critical role land-use systems can play in stabilizing CO_2 concentration in the atmosphere. However, efforts to consider land-based mitigation projects are constrained by uncertainties and by limitations of methodology and data. Any climate change mitigation project that is part of land-based projects requires carbon inventory. Large spatial and temporal variations in the stocks, accumulation rates and losses of the five carbon pools and even within a single vegetation type such as an evergreen forest, grassland and eucalyptus or pines plantation complicate the methods of estimation of carbon benefits. Mitigating climate change through land-use systems requires reduction of emissions from land-use sectors, removal of CO_2 from atmosphere and storing it in vegetation and soil and using biofuels to replace fossil fuels. Such carbon benefits should be permanent and sustainable, but they could be lost by clearing forests, burning biomass, land-use change or even simply disturbing the topsoil. Further, protection of forests in one area could lead to loss of forests and leakage of carbon benefits in another area. Addressing such issues requires methodological and accounting approaches, and a carbon inventory is at the heart of these approaches.

Thirdly, commercial roundwood production as well as community forestry projects require estimation of carbon, particularly in biomass and growth rates of timber and fuelwood production. Finally, grassland improvement, watershed development, land reclamation and halting desertification programmes require carbon inventory.

From a teaching and research perspective, the need for a handbook designed for academic courses and research projects has been obvious during our research and teaching career. This applies to projects aimed at land-use change, vegetation assessments and biomass inventory, where limitations of time, money and experience exist. With this handbook, we hope to help teachers, students, researchers as well as project developers in this process. The aim is also to give adequate background to the methods that can act as a basis for teaching and application for land-use and vegetation inventories.

There are textbooks for professionals, for example, on forest inventory or on soil chemistry. We always wondered why manuals of simple carbon inventory methods are not available despite the need for such practical guidance for a large variety of institutions and individuals. The manual should provide step-by-step guidance on methods of estimating carbon stocks and rates of change to project developers, reviewers, evaluators and, most important, to the beneficiaries, such as industry plantation managers requiring roundwood as raw material, rural communities requiring fuelwood or timber and climate change mitigation project stakeholders.

This handbook is prepared to provide a simple approach to carbon inventory from the perspective of the user covering all aspects of the inventory process at all stages of project conceptualization, planning, proposal development, appraisal, implementation, monitoring and evaluation. The approaches, methods and steps could be applied equally to generating values of carbon stocks and rate of change for carbon emissions and removals for national greenhouse gas inventory estimation as a supplement to the IPCC GHG Inventory Guidelines.

Carbon inventory methods are dynamic and new techniques are being developed and made available. For example, application of remote sensing techniques could become more cost-effective and user-friendly for large-scale applications with technological developments. But traditional methods are robust and simple to adopt and will always be in use. Very often, default data are used in carbon inventory preparation for stock and rate of growth values of different carbon pools. We have attempted to provide valuable sources of default data and approaches to selection and validation of data from different sources.

We were in a dilemma when it came to deciding between the levels of simplicity and detail and hope we have achieved a right balance for the users. We have resisted the temptation to increase the complexity and length of the handbook and have stayed within reasonable limits. For convenience of the user – but at the cost of being repetitious in places – we have attempted to make some chapters self-contained; these are therefore "stand-alone" and can be read independently of the rest of the book. There is a need to change the perception that carbon inventory in land-use systems is contentious and complex and involves large uncertainties. We hope this handbook will go a long way in helping all the potential users in making carbon inventories for different programmes, mechanisms and end uses.

N.H. Ravindranath (R)
Madelene Ostwald (O)

Acknowledgements

This handbook is a result of my long-term collaboration with a large number of authors of several reports of the Intergovernmental Panel on Climate Change, and I cannot risk taking any names, since there are so many of them and I am sure to miss a few. In addition, my work in many developing countries on greenhouse gas (GHG) inventory estimation and carbon mitigation potential assessment in land-use sectors has given me greater insights into the need for a handbook. I am grateful to many of them: the number of such countries is again very large. My effort in preparing a guideline for UNDP-GEF on a manual for carbon inventory also gave me the idea to develop a handbook, for which I am grateful to Richard Hosier and Bo Lim. Further, I and Bo Lim had realized the need for a Handbook during our work for IPCC Report on Land-Use, Land Use Change and Forestry (R).

This book would not have been possible without the dedicated support of many of our colleagues: first, Indu Murthy, who worked so devotedly at various stages of the preparation of this handbook; second, Niranjan Joshi, who gave incisive comments on the draft of this handbook; third, Yateendra Joshi, who was a friendly and efficient copy editor; and, finally, K.S. Murali, R. Chaturvedi, Matilda Palm, Eskil Mattsson, Elisabeth Simelton, Jonas Ardö, G. Chaya, R. Tiwari and G.V. Suresh, who gave comments and assisted us in the preparation of the handbook (R & O).

We also acknowledge the support provided by the Ministry of Environment and Forests, Government of India; the Centre for Climate Science and Policy Research, Linköping University, Department of Earth Sciences, Göteborg University, Sweden and Swedish Energy Agency.

I thank Shailaja for her continued support in the preparation of all my books, which, over the years, has involved long absences from home and negligence of many things that matter in life. I am sure she will continue to support my pursuit of writing more books that are taking shape in my mind (R). I thank Ola, Vilgot and Lisen and the rest of the big family for encouragement and help while working on this book (O).

We acknowledge the support provided by UNDP, in purchasing copies of the handbook, which will go a long way in promoting the use of this handbook in different parts of the world for preparing GHG inventories and developing land-based mitigation projects. Thanks to UNDP, this handbook will be used in training and capacity-building programmes on GHG inventory in land-use sectors. We thank, in particular, Bo Lim, who encouraged and supported work in many developing countries on GHG inventories and carbon mitigation potential assessment in land-use sectors.

N.H. Ravindranath (R)
Madelene Ostwald (O)

Contents

Chapter 1
Introduction

Global environmental problems such as climate change, tropical deforestation, loss of biodiversity and desertification are receiving serious attention of all stakeholders including scientists, citizens and policymakers. Interestingly, all these environmental issues are linked to land-use systems. Climate change and its manifestations, particularly rising temperatures, changing precipitation patterns and sea level rise (IPCC 2007a), are of global environmental concern and have the potential to impact most natural ecosystems (such as forests, grasslands and wetlands) and socio-economic systems (such as food production, fisheries and coastal settlements) in all countries. Under extreme conditions, the impacts are likely to be catastrophic to human survival and may lead to irreversible loss of natural ecosystems. Climate change in the long term is projected to adversely affect supply of fresh water, production of food and forest products and ultimately economic development of both industrialized and developing countries (IPCC 2007b).

Concentrations of greenhouse gases (GHGs) in the atmosphere and their positive radiative forcing have increased since the Industrial Revolution, and particularly faster in the last 30–50 years. Among the GHGs, carbon dioxide (CO_2) is the most dominant, accounting for nearly 77% of the global total CO_2 equivalent greenhouse gas emissions (IPCC 2007c). In terms of radiative forcing by GHGs emitted as a result of human activities, CO_2 accounts for 56% of the total global warming potential of GHGs (IPCC 2007a). The concentration of CO_2 in the atmosphere increased from a pre-industrial level of 279 parts per million by volume (ppmv) to 379 ppmv in 2005. An unprecedented increase of 36% in CO_2 concentration was observed between 1750 and 2005.

Emissions of CO_2, methane and nitrous oxide have increased in the last 25 years (Fig. 1.1a). In 2004, CO_2 emissions from fossil fuels dominated, accounting for 57% of the total global CO_2 equivalent GHG emissions (Fig. 1.1b) whereas those from deforestation, decay of biomass and peat accounted for about 19% (Fig. 1.1b) or 9.5 Gt (gigatonnes or billion tonnes). The CO_2 emissions from land-use sectors have increased from 6.35 Gt in 1970, an increase of 0.126 $GtCO_2$ annually, to 9.5 Gt in 2004. Over the 19th century and much of the 20th century, the terrestrial biosphere has been a net source of atmospheric CO_2 (IPCC 2001a).

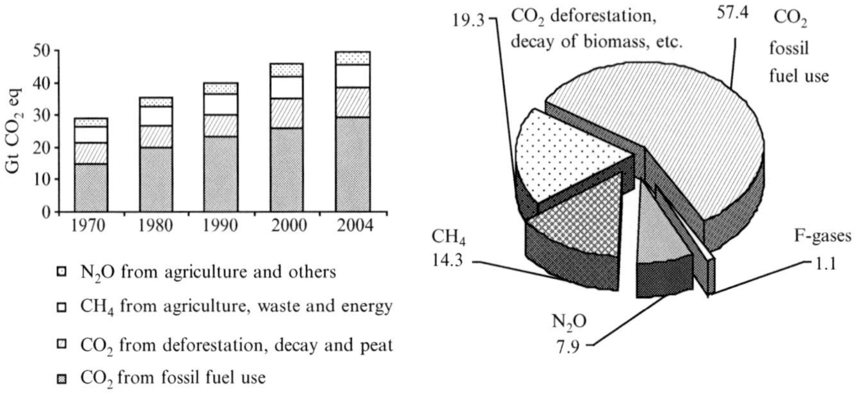

Fig. 1.1 (a) Trends in emissions of global anthropogenic greenhouse gases, 1970–2004, (b) share of different greenhouse gases (IPCC 2007c)

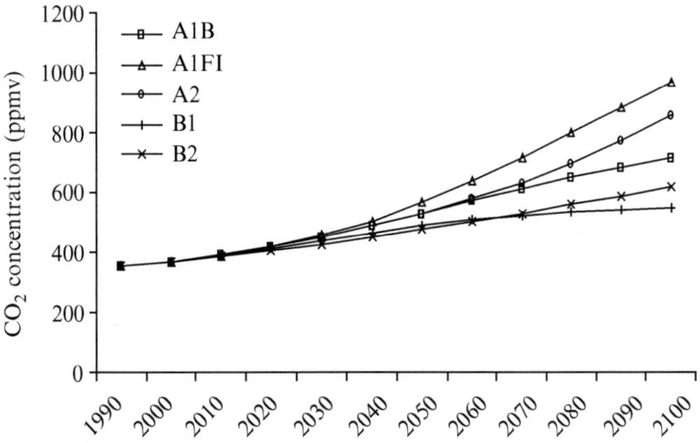

Fig. 1.2 Projected atmospheric concentrations of CO_2 under SRES marker scenarios (see Nakicenovic et al. 2000 for details of the scenarios)

IPCC Special Report on Emission Scenarios (SRES) (Nakicenovic et al. 2000) projects an increase in the atmospheric concentration of CO_2 under all the six illustrative SRES scenarios (Fig. 1.2). The projected concentrations of CO_2, the dominant anthropogenic GHG, in 2100 are likely to range from 540 to 970 ppmv, compared to 279 ppmv in the pre-industrial era (IPCC 2001a).

According to various SRES scenarios, a global warming by 1.8–4.0°C is projected by 2100 with land surface warmer than oceans, along with regional changes in precipitation and sea level rise (IPCC 2007a). Scientific evidence suggests that the projected climate change is likely to lead to increased water scarcity, droughts and high rainfall events, loss of biodiversity, shifts in forest types and reduction in food production in dry tropics with increased risk of hunger and flooding due to sea level rise. Some of the impacts such as loss of biodiversity and wetlands are irreversible (IPCC 2007b).

Mitigation and adaptation are the two approaches adopted by global scientific and policy community to address climate change. Mitigation is defined as the anthropogenic intervention to reduce the sources or enhance sinks of GHGs. Mitigation measures contribute to stabilization of atmospheric concentrations of GHGs, particularly CO_2, at lower levels than projected, thereby reducing the magnitude and rate of climate change. Adaptation is adjustment in natural or human systems in response to actual or expected climatic stimuli and their impacts on natural and socio-economic systems. Adaptation is a necessary strategy to complement mitigation.

Addressing climate change requires scientific, technical and economic assessment of: (i) inventory of CO_2 and other GHG emissions and removals from different sectors; (ii) contribution of different countries to the total global CO_2 emissions and rise in GHG concentration, particularly that of CO_2, in the atmosphere; (iii) different mitigation options and their mitigation potential, costs and benefits; and (iv) development, implementation and monitoring of mitigation projects, strategies and policies, particularly with reference to CO_2. Mitigation opportunities exist largely in energy and land-use sectors. Land-use sector, the focus of this handbook, is critical to addressing climate change concerns through both mitigation and adaptation. Mitigation activities are likely to be implemented in most countries under diverse physical, biological and socio-economic situations, requiring diverse methods for estimation and monitoring of carbon sequestered and CO_2 emissions avoided under different mitigation programmes and projects. This handbook focuses on such methods and guidelines for different land-use sectors as well as methods of carbon inventory for dominant land-use categories, namely forest land, cropland and grassland. Further, the handbook attempts to provide reliable, cost-effective, transparent and comparable methods for inventory of national CO_2 emissions and removals as well as for monitoring carbon stock changes in carbon mitigation programmes and projects.

1.1 Carbon Dioxide Emissions and Removals from Land-Use Sectors

Carbon dioxide is the dominant GHG from land-use sector, particularly forests and grasslands. Forests accounted for about 30%, or 3,952 million hectares (Mha), of the global geographic area, followed by grasslands (about 27%), in 2005. Carbon dioxide emissions from land use and land-use change sector were estimated at 1.6 GtC (ranging from 0.5 to 2.7) annually during the 1990s (IPCC 2007a). Further, the net emissions from land use, land-use change and forest (LULUCF) sector in the tropics at 1.1 ± 0.27 GtC/year (IPCC 2007c), are dominated by conversion of forests (tropical deforestation), followed by loss of soil, forest degradation due to non-sustainable logging and fuelwood collection and forest fire. Estimates of CO_2 emissions therefore range from 0.5 to 2.7 GtC; this wide range highlights the high uncertainty in estimates of CO_2 emissions from land-use sectors. The high uncertainty is due to methodological and data-related issues. Interestingly, GHG inventory estimates show that the LULUCF sector is a sink of carbon in some countries such as Argentina and Zimbabwe and a source in others such as Indonesia and Brazil (www.unfccc.int). IPCC has developed and recommended methods and guidelines for CO_2 as well as other GHG emissions inventories in land-use sectors such as forests, cropland, grassland and wetlands (IPCC 1996, 2006). Carbon inventory in land-use sectors is characterized by high uncertainty in the estimates from the different inventory methods as well as input data.

1.2 Mitigation Opportunities and Potential in Land-Use Sectors

During 2000–2005, the annual net loss of forest area was estimated at approximately 9.39 Mha (FAO 2006). However, the gross annual deforestation rate was approximately 13 Mha globally during the same period. Deforestation is a major source of CO_2 emissions from the LULUCF sector. Further, forests are subjected to disturbances such as forest fires and pests or climatic events of drought and floods, and these disturbances affect about 100 Mha of forests annually. More than 750 Mha of forest land was estimated to have been cleared in developing countries (Houghton 1993; Dale et al. 1993), with more than 90% of the cleared land being managed inefficiently or used for marginal agriculture (Lugo and Brown 1993). Area under forests in the developing countries, largely tropical forests, declined by 200 Mha during the period 1980–2000. The total area under forest plantations was about 140 Mha in 2005, with an annual rate of afforestation of about 2.8 Mha (FAO 2006). Therefore, the most obvious primary mitigation opportunities relate to slowing or halting deforestation and forest degradation, reforestation or afforestation of the deforested or other degraded lands to sequester carbon from the atmosphere and bioenergy projects to replace fossil fuels.

In addition to mitigation opportunities in the forest sector, significant additional mitigation opportunities exist in grassland reclamation and management as well as agricultural sector activities such as agroforestry and land reclamation.

The IPCC Third Assessment Report (Kauppi et al. 2001) concluded that the forest sector can sequester 5.38 $GtCO_2$ per year on an average until 2050, whereas an earlier IPCC report (Watson et al. 2000) had quoted an even higher technical mitigation potential of 11.67 $GtCO_2$ per year. According to IPCC (2007c), the economic mitigation potential in the forest sector ranges from 2.7 to 13.8 $GtCO_2$ annually by 2030 at less than US$100 per tCO_2.

The annual potential for increasing carbon sinks in agriculture sector is estimated to be in the range of 2.9–8.8 $GtCO_2$ (Cole et al. 1996). According to IPCC (2007c), carbon sequestration or mitigation potential in the agriculture sector is 3.87 $GtCO_2$ annually at less than US$100 per tCO_2. Further, the most prominent mitigation opportunity in the agriculture sector relates to carbon sink enhancement through sequestration of carbon in soil though improved cropland and grazing land management, requiring improved estimates.

Thus, carbon mitigation potential in agriculture and forest sector together, excluding bioenergy, is in the range of 6.57–17.6 $GtCO_2$ annually up to 2030 at less than US$100 per tCO_2 (IPCC 2007c). However, mitigation potential estimates in agriculture and forest sectors are characterized by a wide range and high uncertainty.

1.3 Linkages Between Mitigation and Adaptation

Ecological systems such as forests are intrinsically dynamic and are constantly influenced by changes in climate. The IPCC concluded that species composition and dominance could be altered, resulting in ecosystem changes, and populations of many species are at a greater risk due to climate change (IPCC 2001b, 2007b). Forests, grassland and other natural ecosystems are already subjected to socio-economic pressures leading to fragmentation, land conversion and over extraction, all contributing to degradation of ecosystems, biodiversity and carbon sink. Although there are uncertainties about the projected impacts of climate change on forests and other natural ecosystems, evidence that climate change coupled with socio-economic and land-use pressures is likely to adversely impact forests, carbon sinks, biomass productivity and carbon uptake rates continues to mount (Ravindranath and Sathaye 2002; IPCC 2007b). Thus, climate change is likely to impact carbon stocks as well as the mitigation potential in forests, plantations, grasslands and agroforestry systems. This factor may have to be considered in making projections of carbon stocks under mitigation projects in the land-use sectors.

1.4 Why Carbon Inventory?

Carbon inventory involves estimation of stocks and fluxes of carbon from different land-use systems in a given area over a given period and under a given management system. Further, carbon inventory is often referred to as the process of making such estimations. Carbon in land area consists of biomass and the soil carbon pools. The biomass pool includes living above-ground and below-ground biomass, litter and deadwood. The two broad methods for carbon inventory are carbon 'Gain–Loss' and 'Stock–Difference' (IPCC 2006). Carbon inventory is expressed as tonnes of CO_2 emission or removal per hectare or at a project or national level over 1 year. It can also be expressed as changes in carbon stocks in tonnes of carbon per hectare or at a project or national level over a period of time. Net carbon emission indicates loss of CO_2 from biomass and soil to the atmosphere through decomposition or combustion. Net carbon removal or sequestration indicates net CO_2 uptake and storage in biomass and soil.

Carbon inventory methods and guidelines are required for estimation of emissions or removal of CO_2 from biomass and soil or changes in carbon stocks from a given land-use system at a project or national level, resulting from human interventions such as land-use change, extraction of biomass, burning of biomass, soil disturbance leading to oxidation of soil organic matter, afforestation, reforestation, forest conservation and other management activities. Carbon inventory is required for activities related to climate change in land-use sectors as well as other forest conservation and development projects such as roundwood production and forest conservation not directly aimed at climate change mitigation. The broad categories of programmes requiring carbon inventory are discussed in the following sections.

1.4.1 Carbon Inventory for National Greenhouse Gas Inventory

All the Signatory Parties of the United Nations Framework Convention on Climate Change (UNFCCC) are required to prepare national greenhouse gas inventories periodically and report them to the UNFCCC. Greenhouse gas inventory includes estimation of emissions and removals of GHGs such as CO_2, CH_4 and N_2O. Under the UNFCCC, Annex I or industrialized countries are expected to estimate and report the GHG emissions inventory annually whereas non-Annex I or developing countries are expected to submit the inventory only periodically and the interval between successive inventories is flexible (about 3–5 years). Most of the developing countries have already estimated and reported their GHG inventory as a component of their National Communication to UNFCCC (www.unfccc.int). The IPCC provides broad guidelines for estimating CO_2 emissions and removals from different land-use sectors at the national level (IPCC 1996, 2006). However, the guidelines do not provide such details as sampling methods, measurement techniques, models for projections and for periodic monitoring of carbon stock changes and calculation procedures to convert the indicator parameters measured in the field and laboratory to tonnes of carbon stock changes per hectare.

1.4.2 Carbon Inventory for Climate Change Mitigation Projects or Programmes

Land-use sectors have been recognized as critical to addressing climate change concerns. Mitigation of climate change through land-based activities has been a contentious issue in global negotiations under the UNFCCC and the Kyoto Protocol because of several methodological issues related to measurement, monitoring, reporting and verification of carbon benefits (Ravindranath and Sathaye 2002). Carbon inventory for mitigation projects requires methods for estimating carbon stocks and changes due to project activities for selected periods at the project concept formulation, proposal development, project implementation and monitoring stages. Methods are required for the baseline (without project) and mitigation scenarios. Mitigating climate change through land-use sectors involves reducing CO_2 emissions or enhancing carbon sinks in biomass, soil and wood products. Reducing deforestation, sustainable forest management, afforestation, reforestation, agroforestry, urban forestry, shelter belts, grassland management and substitution of fossil fuels with bioenergy are examples of mitigation opportunities in land-use sectors.

A carbon inventory for climate change mitigation projects and programmes in the land-use sector involves monitoring and estimation of additional or incremental carbon emission avoided or carbon sequestered due to implementation of a project activity or a programme or a policy initiative. Carbon inventory leads to estimation of tonnes of CO_2 emissions avoided by halting deforestation and fossil fuel substitution or, alternatively, estimation of tonnes of carbon sequestered in biomass and soil as a result of enhancement of sinks through afforestation, reforestation and grassland reclamation activities. The focus of this handbook is on measurement and monitoring of carbon stock changes in biomass and soil in land-based projects and programmes. The methods and guidelines can also be used for estimation and monitoring of carbon stock changes due to avoided deforestation activities.

1.4.3 Carbon Inventory for Clean Development Mechanism Projects

Clean Development Mechanism (CDM) is one of the project-based mechanisms of the Kyoto Protocol to address climate change. The mechanism aims at reducing GHG emissions or enhancing carbon sinks in developing countries and in turn assists Annex I Parties or industrialized countries in meeting their greenhouse gas emission reduction commitments, promoting sustainable development in developing countries at the same time. Afforestation and reforestation are the two project activities permitted under CDM, and these projects aim at increasing the carbon stocks in the project area additional to the carbon stock changes that would have occurred under the baseline (without project) scenario. Such projects are required under the CDM to make estimations and projections of carbon stocks at the project development phase itself, using a methodology approved by the CDM Executive Board. Further, the project proposal should describe the methods to be adopted for monitoring the carbon stock changes as well as actual periodic monitoring and verification of the carbon stocks.

1.4.4 Carbon Inventory for Projects Under the Global Environment Facility

Global Environment Facility (GEF) is one of the global mechanisms aimed at promoting environmentally sustainable technologies, policies, measures and institutional capacity for addressing the goals of the Climate Convention as well as the Convention on Biological Diversity and other global environmental concerns. The GEF has been adopted by the UNFCCC as a financing mechanism aimed at addressing climate change. The facility operates on the principle of providing incremental cost to activities under approved Operational Programmes (OPs) for providing global environmental benefits and has 15 such programmes, of which OP 12 and OP 15 require carbon inventory from land-use projects.

OP 12 "Integrated Ecosystem Management" is a multifocal area programme. The projects under OP 12 aim at multiple benefits including enhancing the storage of carbon stock in terrestrial and aquatic ecosystems and conservation of biodiversity.

OP 15 "Sustainable Land Management" aims at multiple objectives such as prevention of land degradation and global environmental benefits including reduction of CO_2 emissions, increased carbon sequestration and conservation of biodiversity.

One of the key global environmental objectives of OP 12 and OP 15 is to reduce CO_2 emissions and enhance carbon stocks. Carbon inventory is required to estimate the impact of GEF project activities on carbon stocks in the land area covered under the project. This requires estimation of carbon stocks in biomass and soil in the baseline (or without project) and project scenarios and finally the incremental carbon stock gain due to the project activities. The methods and guideline provided in this handbook could be adopted for Operational Programmes under the GEF.

1.4.5 Carbon Inventory for Forest, Grassland and Agroforestry Development Projects

A large number of countries are implementing forest and grassland conservation, roundwood production and agroforestry development programmes and projects such as

- Forest and biodiversity conservation
- Community forestry and industrial roundwood plantations
- Agroforestry, urban forestry and shelter belts
- Grassland reclamation or development

Carbon inventory, which includes estimation of stocks or rates of change in roundwood biomass and soil carbon, is normally required for any afforestation, reforestation, forest conservation and land reclamation programmes or projects. These projects are planned and implemented for non-climate-related objectives at the national and local level and aim at conserving forests and biodiversity, enhancing biomass supply for households and industries, increasing agricultural productivity, improving the management of grasslands and reclaiming land through soil and moisture conservation and halting desertification. Carbon inventory is required to estimate the biomass or roundwood (fuelwood or commercial roundwood) or grass production, and increase in soil organic matter. Further, carbon inventory can be used to assess the impact of a land development project on employment, income and livelihoods through enhanced biomass production and supply from projects on forest, grassland and cropland systems. The biomass stock or production and soil organic matter status estimation and monitoring requires carbon inventory methods throughout the life of a project: at the project formulation stage, during the project and at the end.

1.5 Carbon Inventory Methods and Guidelines

A few manuals or guidelines on carbon inventory methods are available, including, for example, the IPCC guidelines for estimating national GHG or carbon inventory for land-use change and forest sector (IPCC 1996) and for agriculture, forest and other land uses (IPCC 2006) as well the IPCC Good Practice Guidance for LULUCF sector (IPCC 2003). Examples of project-level carbon inventory guidelines are provided by Winrock (www.winrock.org), FAO (www.fao.org), CIFOR (www.cifor.org), the Biomass Assessment Handbook (Rossillo-Calle et al. 2006), Forestry Handbook (Wenger 1984) and Forest Inventory (Kangas and Maltamo 2006).

Adoption of carbon inventory methods and guidelines should lead to accurate, reliable and cost-effective estimates of carbon stocks and changes for a given land-use system and period. Biological systems such as forests and grasslands are characterized by wide variation across locations, management systems and periods even for a given land-use system within a region such as a eucalyptus plantation. Further, it is also subject to high uncertainty due to diverse methods, data limitations and high cost. This handbook on carbon inventory methods and guidelines covers the following topics:

- Description of methods for estimation and monitoring of carbon stocks or emissions and removals
- Criteria for selecting the methods
- Sampling methods and procedures
- Field, laboratory and remote sensing tools and techniques
- Field and laboratory measurement and recording protocols
- Modelling and projections
- Calculation and estimation procedures
- Uncertainty estimation

The need for methods and guidelines for carbon inventory for climate change mitigation projects, national greenhouse gas inventories and forest, cropland and grassland development projects is highlighted in the preceding sections.

1.6 Purpose, Organization and Target Groups for the Handbook

A carbon mitigation project developer, a monitoring expert or a GHG inventory compiler requires access to a manual of guidelines and methods that provides practical guidance using a cookbook approach, describing reliable and cost-effective methods, sampling procedures, field and laboratory measurement techniques, calculation procedures, modelling and reporting protocols and so on. However, no single existing manual or book provides all the guidelines or information required for carbon inventory. The existing manuals and books do not describe different methods required for carbon inventory at project development, implementation and monitoring stages, nor do they adequately cover the modelling aspects of projecting carbon stock changes. Further, the guidance available in existing books and manuals is not project specific, and applications of remote sensing and GIS techniques are not described adequately. The means of access to and sources of data for carbon inventory are rarely addressed, and calculation procedures are often not included. A user may have to refer to several books, manuals and reports for getting complete information on carbon inventory. Despite the importance of carbon inventory for land-use sector activities or projects, no comprehensive handbook or a set of guidelines exists to assist those engaged in carbon mitigation project development, implementation and monitoring or national greenhouse gas inventory agencies or forest or grassland or agricultural development project managers.

This handbook is aimed at providing practical guidance in a user-friendly way on the methods for carbon inventory for a variety of projects and purposes.

- Carbon mitigation project development, implementation and monitoring in forest, agriculture and grassland sectors (e.g. projects under the Clean Development Mechanism or Global Environment Facility or those under the World Bank's Biocarbon Project)
- National greenhouse gas inventory in LULUCF or forest, agriculture, grassland and other land-use categories
- Forest, grassland and agroforestry development projects targeted not at climate change mitigation but at conserving biodiversity, improving soil fertility or conserving soil and water
- Commercial forestry and plantations meant for roundwood production
- Community forestry programmes requiring information on biomass or fuelwood growth rates or production
- Modelling of projections of carbon stock changes

Carbon inventories are prepared by a diversity of institutions, managers, consultants, experts and individuals, from academic and non-academic sectors. The target groups for this carbon inventory handbook include the following:

- *Universities and research institutions* with forestry, land-use and climate change disciplines
- *Consultancy agencies* involved in developing or monitoring forestry, agriculture and grassland-related projects
- *Non-governmental organizations (NGOs)* involved in forest, agriculture and grassland mitigation project development and monitoring
- *Donor agencies* such as the World Bank and other multilateral institutions funding carbon mitigation projects
- *Individual climate mitigation project developers* in the developing and industrialized countries
- *National greenhouse gas inventory agencies, inventory compilers and reviewers*

- *Developers and managers* involved in roundwood production and land reclamation projects
- *United Nations agencies and mechanisms* such as the Clean Development Mechanism institutions, Operating Entities under CDM, carbon certification agencies and the Global Environmental Facility
- *Forest service or departments and community forestry organizations*

The practical guidance offered in the handbook is supplemented with explanations of the scientific basis for the recommended practices, methods, field and laboratory measurement and monitoring techniques, models and calculation procedures.

Chapter 2
Global Carbon Cycle, Carbon Dioxide Emissions and Mitigation

The carbon cycle is one of the biogeochemical cycles and describes the movement of carbon, in its many forms, within the biosphere, atmosphere, oceans and geosphere. The global carbon cycle involves the earth's atmosphere, oceans, vegetation and soils of the terrestrial ecosystem and fossil fuels. Carbon in the form of inorganic and organic compounds, notably carbon dioxide (CO_2), is cycled between different components of a system. For example, green plants absorb CO_2 from the atmosphere during photosynthesis, also called primary production, and release CO_2 back into the atmosphere during respiration. Another channel of exchange of CO_2 is between the oceans and the atmosphere: CO_2 dissolved in the oceans is used by marine biota in photosynthesis.

Two important anthropogenic processes that contribute CO_2 to the atmosphere are burning of fossil fuels and changes in land use. Fossil fuels, namely coal, oil and natural gas, are burnt in industries, power plants and automobiles. Land use is a broad term, which encompasses a host of essentially human-induced activities including conversion of natural ecosystems such as forests and grasslands to managed systems such as cropland, grazing land and settlements. Land conversion and other human activities such as extraction and burning of biomass and livestock grazing lead to soil degradation and emission of carbon contained in biomass and in soil to the atmosphere: CO_2 emissions from the biosphere to the atmosphere result mainly from burning and decomposition of organic matter.

2.1 Carbon Stocks and Fluxes

The contribution of different components to the global carbon cycle covering a recent period (2000–2005) is shown in Fig. 2.1. The stock of carbon in the atmosphere is increasing due to anthropogenic activities such as fossil fuel combustion, cement production and land-use change. The terrestrial pool has an annual uptake of carbon from the atmosphere, described as gross primary productivity (GPP), amounting to about 120 GtC; a little over half of which is returned to the atmosphere through

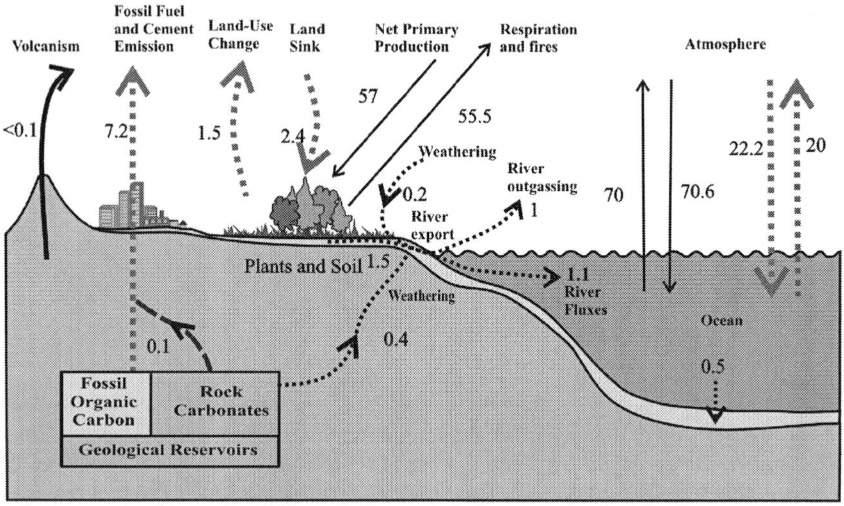

Fig. 2.1 Global carbon cycle (SCOPE 2006)

plant respiration, leaving 57 GtC in the terrestrial sink as net primary productivity (NPP). Fossil fuel combustion and cement production contribute about 7.2 GtC annually and land-use contributes about 1.5 GtC annually (SCOPE 2006).

2.2 Anthropogenic CO_2 Emissions

The emissions of CO_2 have increased continuously during the recent decades (Fig. 1.1). Global average annual CO_2 emissions from fossil fuels increased from 5.4 ± 0.3 GtC in the 1980s to 7.2 ± 0.3 GtC during 2000–2005 (Table 2.1); those from land-use change increased from 1.4 GtC during the 1980s to 1.6 GtC during the 1990s. It is important to note that the range of estimated emissions from land-use change is very wide, as much as 0.5–2.7 GtC for the 1990s, indicating the high level of uncertainty.

Since the 1980s, natural processes of CO_2 uptake by the terrestrial biosphere (the residual land sink) and by the oceans have removed about 50% of anthropogenic emissions from fossil fuel combustion and land-use change. Carbon uptake and storage in the terrestrial biosphere arise from the net difference between uptake due to growth of vegetation that is changes in afforestation and reforestation leading to carbon sequestration, and emissions due to heterotrophic respiration, harvest, deforestation, fire, pollution and other disturbances affecting biomass and soils (IPCC 2007a).

Table 2.1 Global annual carbon budget (GtC) (IPCC 2007a)

Source	1980s	1990s	2000–2005
Atmospheric increase	3.3 ± 0.1	3.2 ± 0.1	4.1 ± 0.1
Fossil carbon dioxide emissions	5.4 ± 0.3	6.4 ± 0.4	7.2 ± 0.3
Net ocean-to-atmosphere flux	−1.8 ± 0.8	−2.2 ± 0.4	−2.2 ± 0.5
Net land-to-atmosphere flux	−0.3 ± 0.9	−1.0 ± 0.6	−0.9 ± 0.6
Partitioned as follows			
Land-use change flux	1.4 [0.4–2.3]	1.6 [0.5–2.7]	NA
Residual land sink	−1.7 [−3.4 to 0.2]	−2.6 [−4.3 to −0.9]	NA

1 GtC = 1 billion tonnes of carbon = 10^9 t of carbon = 3.666 × 10^9 t of CO_2

2.3 CO_2 Concentration in the Atmosphere

Natural as well as human-induced activities contribute to changes in the concentration of CO_2 in the atmosphere. Studies and analyses of air bubbles trapped within ice cores from Greenland and Antarctica have provided researchers with evidence of variation in atmospheric CO_2 concentrations since the end of the last glacial maximum. Starting at about 200 parts per million by volume (ppmv), the concentration rose to 275–285 ppmv before the start of the Industrial Revolution and reached 366 ppmv in 1998 (Keeling and Whorf 1999), a level unprecedented during the past 420,000 years (Petit et al. 1999).

According to the latest estimates, the concentration of atmospheric CO_2 increased from a pre-industrial level of about 280 to 380 ppmv in 2006. Whereas over 8,000 years prior to industrialization the concentration had increased only by 20 ppmv, since 1750 it has risen by nearly 100 ppmv. The annual CO_2 growth rate during the last 10 years has been higher (1.9 ppmv was the annual average during 1995–2005) than it had ever been since continuous direct atmospheric measurements began (1.4 ppmv annually during 1960–2005). Use of fossil fuels and effects of land use and land-use change on plant and soil carbon are the primary sources of increased atmospheric CO_2, their estimated respective shares since 1750 being 65% and 35% (IPCC 2007a). Depending on the scenario, by 2100 the concentration is projected to increase to 540–970 ppmv (Fig. 1.2), which is 90–250% above that in 1750. It is very important to understand the contribution of different countries and sectors to the increasing CO_2 concentration in the atmosphere although the current estimates of CO_2 emissions from land-use sectors are characterized by high uncertainty.

2.4 Carbon Stocks in Vegetation and Soils of Different Biomes

The stocks of carbon in different biomes and the carbon flow pathways are depicted in Fig. 2.2. It can be observed that forest biome dominates the carbon stocks and has more carbon than the total carbon stock in the atmosphere. The emissions of CO_2 follow a number of pathways (respiration, decomposition and combustion), and in addition carbon sequestration occurs in forests, grasslands and croplands.

Globally, out of the total 151 million square kilometres of terrestrial ecosystem area, forests account for approximately 27%, followed by tropical savannas and grasslands, which together account for about 23% and croplands, about 10% (Table 2.2). At the global level, the quantity of carbon stored in the terrestrial ecosystem is estimated at 2,477 GtC; soil accounts for approximately 81% of it and vegetation accounts for the rest (Table 2.2).

Among the different biomes, only tropical forests hold as much carbon in vegetation as in soils whereas in all the other biomes soils claim a much larger share. Carbon stock in vegetation is insignificant compared to that in soils in temperate grasslands, deserts, semi-deserts, wetlands and croplands. Thus, it is

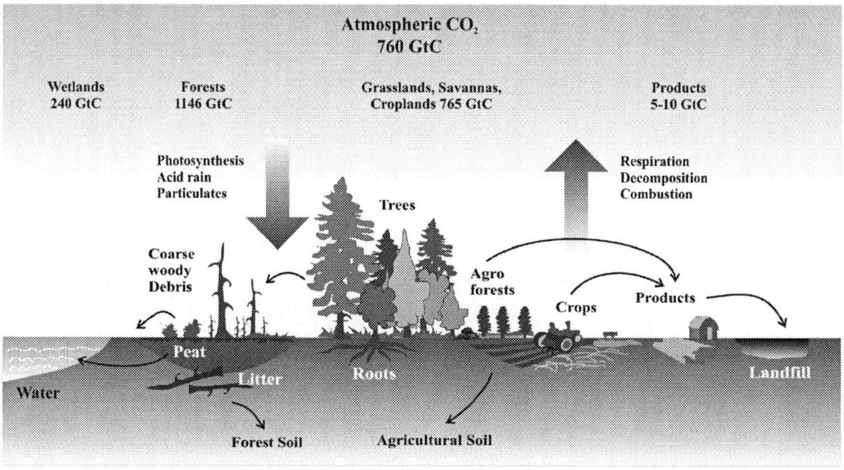

Fig. 2.2 Stocks of carbon, different carbon pools and carbon flow pathways (Kauppi et al. 2001)

Table 2.2 Global carbon stocks in vegetation and top 1 m of soils (Modified from WBGU 1998 and Watson et al. 2000.)

Biome	Area (10^6 km^2)	Carbon stocks (GtC) Vegetation	Soils	Total
Tropical forests	17.6	212	216	428
Temperate forests	10.4	59	100	159
Boreal forests	13.7	88	471	559
Tropical savannas	22.5	66	264	330
Temperate grasslands	12.5	9	295	304
Deserts and semi-deserts	45.5	8	191	199
Tundra	9.5	6	121	127
Wetlands	3.5	15	225	240
Croplands	16.0	3	128	131
Total	151.2	466 (19%)	2,011 (81%)	2,477 (100%)

important to focus on methods for estimation of soil carbon pools in grasslands, savannas and croplands, whereas methods that cover vegetation as well as soil carbon are required to estimate carbon in forests and agroforestry systems.

2.5 CO_2 Emissions from Land-Use Sectors

Emission of CO_2 from land-use sectors result from changes such as tropical deforestation and conversion of grasslands to cropland, degradation of forests and grasslands, biomass decay and burning. Net cumulative global CO_2 emissions from land-use change during 1850–1998 are estimated at 136 ± 55 GtC; out of this, 87% emissions are from forest areas (Bolin and Sukumar 2000), attributed largely to tropical deforestation. Compared to annual average emissions of 0.9 GtC during 1850–1998, those from land-use change increased to 1.6 GtC (0.5–2.7 GtC) during the 1990s (Table 2.1). Over the last few decades, CO_2 emissions have increased: from approximately 0.74 GtC in 1960, the level went up to about 1.6 GtC in 2005, indicating continuous growth in CO_2 emissions from land-use sectors (Fig. 2.3).

The main source of CO_2 emissions from land-use sectors is tropical deforestation: forest area declined annually by 8.8 Mha during the 1990s; however, according to the latest FAO assessment, the rate came down to 7.9 Mha during 2000–2005, a 10% reduction over that in the 1990s (Table 2.3). Africa and South America dominate, with each accounting for about 4 Mha during those 5 years. According to FAO (2006), ten countries with the largest net annual loss in forest area during 2000–2005 were

- *South America*: Brazil (3.01 Mha), Venezuela (0.28 Mha)
- *Asia*: Indonesia (1.87 Mha), Myanmar (0.46 Mha)
- *Africa*: Sudan (0.59 Mha), Zambia (0.44 Mha), Tanzania (0.41 Mha), Nigeria (0.41 Mha), DR Congo (0.32 Mha) and Zimbabwe (0.31 Mha)

The area under forests in Asia and Europe has increased: among the countries with the largest annual net gain in forest area, China topped the list with 4.05 Mha (FAO 2006).

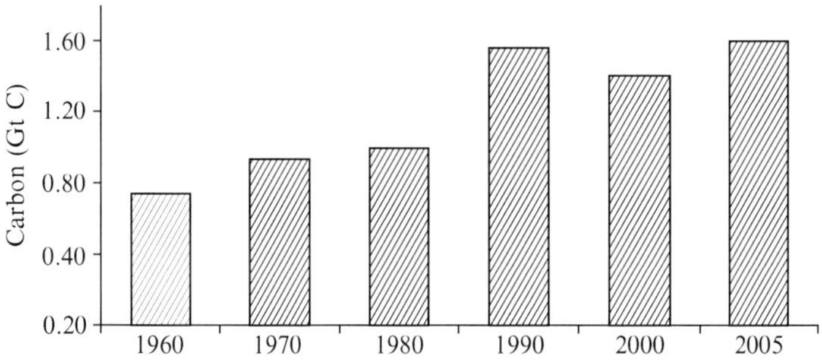

Fig. 2.3 CO_2 emissions from tropical land-use change during 1960–2005 (www.globalcarbonproject.org/budget.htm)

Table 2.3 Changes in forest area, by region (FAO 2006)

	Forest area (1,000 ha)			Annual forest area change (1,000 ha)	
Region	1990	2000	2005	1990–2000	2000–2005
Africa	699.36	655.61	635.41	–4.37	–4.04
Asia	574.49	566.57	571.58	–0.79	1.00
Europe	989.32	998.09	1,001.39	0.87	0.66
North and Central America	710.79	707.51	705.85	–0.32	–0.33
Oceania	212.51	208.04	206.25	–0.44	–0.35
South America	890.82	852.79	831.54	–3.80	–4.25
World	4,077.29	3,988.61	3,952.02	–8.86	–7.31

Future trends in carbon emissions and removals by forests are difficult to predict. Available evidence indicates that for several decades: (i) deforestation in the tropics will continue to be a major source of carbon emissions (Canadell et al. 2004); (ii) carbon sink in countries with economies in transition may decrease as forests mature and may become a source if the growing economies cause increased logging; (iii) carbon stocks in forests of many developed countries will continue to grow, unless they choose to rely more on their forests to meet the demand for timber and land; and (iv) aggressive reforestation programmes in previously deforested countries can produce new carbon sinks (IPCC 2007c).

2.6 Mitigation Potential in the Land-Use Sectors

Forest and agriculture sectors are critical to stabilizing CO_2 concentration in the atmosphere for mitigating climate change because both offer a large mitigation potential, besides providing multiple sustainable development benefits such as biodiversity conservation, protection of watersheds, sustained roundwood supply, increased crop and grass productivity and livelihoods for forest-dependent communities.

2.6.1 Forest Sector

The broad mitigation options in forest sector include (IPCC 2007c):

- Maintaining or increasing forest area
- Maintaining or increasing carbon density at the site level
- Maintaining or increasing carbon density at the landscape level
- Increasing off-site carbon stocks in wood products and enhancing product and fuel substitution

According to the latest assessment by IPCC, the economic mitigation potential is in the range of 1.6–5 GtCO_2 annually during the period up to 2030 at less than US$20 per t$CO_2$; however, at mitigation costs of less than US$100 per t$CO_2$, the potential rises to 2.7–13.8 (IPCC 2007c). It is important to note the wide range in the estimates, reflecting the high uncertainty. Among the mitigation options in the forest sector, avoided deforestation offers maximum potential.

2.6.2 *Agriculture Sector*

In the agriculture sector, the dominant carbon mitigation options include restoration of cultivated organic soils, better management of cropland and grazing land and restoration of degraded lands. The mitigation potential of the sector is dominated by carbon sink enhancement of agricultural soils: the potential of carbon sequestration in soils is estimated to account for 90% of the total mitigation potential of agriculture sector and involves the following activities:

- Restoration of cultivated organic soils (1,260 Mt CO_2)
- Improved cropland management (1,110 Mt CO_2)
- Improved grazing land management (810 Mt CO_2)
- Restoration of degraded lands (690 Mt CO_2)

2.7 Conclusions

An understanding of the carbon cycle, particularly the impacts of human activities on global carbon stocks and flows, is critical to addressing climate change. The dominant human activities relevant to the carbon cycle are: (i) emission of CO_2 from fossil fuel combustion and land-use change; (ii) mitigation through carbon emission reduction (through avoided deforestation); (iii) carbon sequestration in vegetation and soils; and (iv) substitution of fossil fuels with bioenergy. Estimates of annual CO_2 emissions from land-use sectors are characterized by a wide range (0.5–2.7 GtC for the 1990s) and high uncertainty, and so are the estimates of mitigation potential in the forest sector (0.73–3.76 GtC up to 2030). Thus, there is a need for improved methods and better data for reliable estimates of CO_2 emissions from land-use sectors as well as carbon mitigation potential of forest and agricultural sectors. It is also important to note the dominance of soil over vegetation in storing carbon in savannah, grasslands, pastures and cropland. Therefore, methods for estimating soil carbon are critical to these land-use categories and methods for estimating above-ground tree biomass along with soil carbon are critical to forests, plantations and agroforestry systems.

Chapter 3
Categories of Activities, Programmes and Projects Requiring Carbon Inventory

Chapter 1 briefly describes the rationale and need for carbon inventory methods and guidelines and touches upon different programmes, activities and projects requiring carbon inventory. A carbon inventory involves estimation of changes in the stocks of carbon (in biomass and in soil) or its emissions and removal, normally expressed in tonnes of carbon per hectare, for a given land-use system, and time at a project level or national level. Carbon inventory is required for estimating: (i) the contribution of a country to global greenhouse gas (GHG) emission; (ii) carbon mitigation potential of a given project or activity; (iii) production of commercial timber or fuelwood from a plantation forestry project; and (iv) the impact of land reclamation projects on soil fertility.

3.1 National Greenhouse Gas Inventory

Greenhouse gas inventory at the national level includes estimation of emissions and removals of GHGs in agriculture, forest land, grassland and other land-use categories. The key GHGs of concern for land-use sectors are CO_2, nitrous oxide (N_2O) and methane (CH_4). Of these, CO_2 is the dominant GHG, and CO_2 fluxes between the atmosphere and the land are primarily controlled by uptake for plant growth and releases from respiration, decomposition and oxidation of organic matter. Emissions and removals for a given land-use category or subcategory are estimated by using the area under the category or subcategory and changes in stocks of different carbon pools.

$$\text{Annual change in carbon stock (tC/ha)} = \text{Sum of annual changes in different carbon pools} (AGB + BGB + DW + LT + SC)$$

where, C is carbon, AGB is above-ground biomass, BGB is below-ground biomass, DW is deadwood, LT is litter and SC is soil organic carbon.

Preparing a carbon inventory for the national GHG inventory for a given land-use category such as forest land, grassland or cropland involves two approaches (IPCC 1996, 2006):

- *Carbon "Stock–Difference" method*, which requires estimation of stocks of different carbon pools over two periods
- *Carbon "Gain–Loss" method*, which requires estimation of annual gains and losses in carbon stocks

The key steps in estimation of carbon inventory for national GHG inventory are as follows:

Step 1: Estimate the area under a given land-use category and subcategory in a given year as well as area under each land-use category subjected to land-use change

Step 2: Estimate the stocks of carbon in each of the carbon pools at the beginning and end of the period, the difference between the two periods being the net emission or removal

Step 3: Alternatively, estimate the gain in carbon stock for each of the carbon pools due to growth or accumulation and losses from each pool due to harvest and disturbance. Next, estimate the difference between carbon gain and losses as the net emission or removal.

Methods for carbon inventory in forest land, cropland, grassland and other land-use categories include those for estimating the following quantities:

- Area and area changes under each land-use category and subcategory
- Carbon accumulation in above-ground and below-ground biomass or stock changes in biomass pools over time
- Changes in stocks of litter and deadwood or flows into and out of these pools
- Changes in soil organic carbon stock or rate of accumulation
- Losses in carbon held in biomass and in soil due to harvest and disturbance

Table 3.1 Carbon inventory categories for national greenhouse gas inventory according to IPCC (1996) and IPCC (2006)

IPCC 1996	IPCC 2006
5A: Changes in forest and other woody biomass stocks	*Forest land*
	(a) Forest land remaining forest land
	(b) Land converted to forest land
5B: Forest and grassland conversion	*Cropland*
	(a) Cropland remaining cropland
	(b) Land converted to cropland
5C: Abandonment of croplands, pastures, plantation forests, or other managed lands	*Grassland*
	(a) Grassland remaining grassland
	(b) Land converted to grassland
5D: CO_2 emissions and removals from soils	*Wetland*
	(a) Wetland remaining wetland
	(b) Land converted to wetland
	Settlement
	(a) Settlement remaining settlement
	(b) Land converted to settlement

A carbon inventory, as part of estimating national GHG emissions and removals, has to be prepared and reported by Annex-I or industrialized countries and the non-Annex-I or developing countries. IPCC (1996, 2006) provides guidelines for preparing such an inventory for different land-use categories (Table 3.1), and this handbook supplements those guidelines with details and step-by-step procedures and describes, in Chapter 16, how the guidelines can be applied to preparing a national GHG inventory.

3.2 Carbon Inventory for Climate Change Mitigation Projects and Programmes

The mitigation potential of land-use sectors, particularly forests and cropland, has been estimated by several studies at the global, national and, in a few cases, at the project level (IPCC 2007c; Ravindranath and Sathaye 2002). The uncertainty in these estimates is very high due to methodological as well as data-related issues (Chapter 18). Mitigation programmes and projects normally involve estimation of verifiable changes in carbon stocks over a given period, and estimation of the mitigation potential of a project or programme requires measurement, estimation and reporting of carbon stock changes in a given area under the "without-project" or "baseline scenario" – i.e. in absence of a mitigation programme or project activities – as well as carbon stock gain due to implementation of the mitigation programme or project activities under the "mitigation scenario". Land-use-based mitigation activities normally involve conserving the existing carbon stocks or increasing the carbon stocks or density in low-carbon-degraded areas or biomass energy replacing fossil fuels, leading to a net CO_2 emission reduction. Mitigation through conservation of carbon stocks and expansion of carbon sinks due to a programme or project activity could be captured by estimating stock change between two points in time. Methods and guidelines are required for the following components:

- Selection of a specific mitigation programme or project activity.
- Estimation of area under the mitigation activity and marking the boundary.
- Identification of relevant carbon pools for the selected activity.
- Selection of a sampling method and of suitable field and laboratory measurement techniques.
- Measurement or estimation of carbon stocks for the selected carbon pools under the "without-project" (baseline) scenario in the project area.
- Measurement of carbon stock changes for the selected carbon pools under the mitigation scenario.
- Estimation of net mitigation potential or net carbon stock gain.
- Methods for carbon inventory vary with the type of mitigation activity. The carbon pools to be selected for an inventory also depend on the type of mitigation activity. Mitigation options in land-use sectors could be grouped in two ways.

Firstly, the forest sector mitigation opportunities can be broadly categorized into three groups (Brown et al. 1996):

- *Forest carbon conservation and management measures*, which involve reducing deforestation and sustainable forest management
- *Carbon storage management*, which involves afforestation, reforestation, agro-forestry and shelterbelts
- *Fossil fuel substitution and management activities*, which involve bioenergy replacing fossil fuels and sustainable timber products replacing such construction materials as cement, aluminium and steel

Table 3.2 Mitigation activities according to land-use categories and their features

Land-use category	Broad mitigation activity	Examples of mitigation activities or projects	Items in carbon inventory
Forest land	Afforestation	-Community woodlots - Commercial plantations - Fruit orchards	- Estimation of increase in carbon stocks in biomass and in soil - Production of fuelwood, timber, etc.
	Reforestation	- Natural regeneration - Plantations in degraded forest land	- Estimation of carbon stocks in biomass - Production of roundwood and estimation of carbon stocks in soil
	Reducing deforestation	- Protected area - Eco-development - Plantation forestry	- Estimation of emissions avoided - Stocks of carbon conserved in biomass and soil
	Forest management	- Reduced-impact logging - Fire management - Fertilizer application	- Estimation of carbon stocks saved from loss - Increase in biomass stocks
Cropland	Agroforestry	- Row intercropping - Shelterbelts - Watershed management - Zero tillage	- Estimation of carbon stocks in soil - Increase in perennial crop biomass production
Grassland	Improved management	- Grazing management - Improved grassland practices	- Increase in grass production and soil organic carbon
Settlements	Urban forestry	- Parks - Avenue trees - Homestead gardens	- Increase in tree crown - Increase in stocks of carbon in tree biomass
Forest land or cropland or grassland	Bioenergy plantations	- Short-rotation woody plantation	- Estimation of stock change in soil and biomass carbon

Secondly, the mitigation options in forest sector could be grouped into four general categories (IPCC 2007c):

- *Maintaining or increasing the forest area* through the avoidance of deforestation and through afforestation/reforestation
- *Increasing the stand-level carbon density (tonnes of carbon per hectare)* by using planting, site preparation, tree improvement, fertilizers, uneven-aged stand management or other silvicultural techniques that contribute to sustainable forest management
- *Increasing the landscape-level carbon density* by using longer forest rotations, fuel management and protection against fire and insects
- *Increasing carbon stock in wood products and enhancing product substitution* by using forest-derived biomass to substitute products with high fossil fuel requirements and increasing the use of biomass-derived energy to substitute fossil fuels

Mitigation activities could be planned and implemented on different land-use categories. Potential land-use categories for mitigation activities include forest land, cropland, grassland and settlements. The carbon inventory methods are presented in this handbook according to different carbon pools applicable to broad land-use categories and specific mitigation activities. An illustrative set of land-use categories, mitigation activities and broad features of mitigation activities, which determine the methods to be adopted, is presented in Table 3.2.

3.3 Carbon Inventory for Clean Development Mechanism Projects

Clean Development Mechanism (CDM) is one of the mechanisms of the Kyoto Protocol to address climate change. The goal of CDM is to assist non-Annex-I countries in achieving sustainable development and Annex-I countries in achieving compliance with the commitment on Quantified Emissions Limitations and Reductions. The Annex-I countries that have ratified the Kyoto Protocol have to limit their GHG emissions during 2008–2012 to the level agreed to in the Protocol. The design of a land-use or forest-based CDM project consists of many elements that are similar to the design of any kind of land-use or forest-based project, but also has some additional features. These general features include establishing a project boundary and a monitoring area and evaluation of environmental and social impacts apart from estimating the additionality of the carbon stock gained. The unique features of CDM project activities are as follows:

- Only afforestation and reforestation project activities eligible under CDM
- Requires development of "without-project" or "baseline" scenario and estimation of the associated carbon stock changes

- Estimation of carbon stock gained under the project scenario due to the implementation of project activities
- Estimation of leakage, that is carbon dioxide emissions resulting from the mitigation project activities within or outside project area
- Estimation of additionality, which is the additional carbon stock gained due to the project activity over the "without-project" or "baseline" scenario carbon stocks
- Only those methodologies approved by the CDM Executive Board for the afforestation and reforestation projects to be adopted (new methodologies need to be approved before a project is accepted)
- Periodic monitoring and verification of carbon stock changes; methodologies and plan for monitoring project proposal to be provided in the proposal

The methodologies approved by the CDM Executive Board for afforestation and reforestation provide broad guidelines for carbon stock estimation for baseline and mitigation project scenario. These methodologies provide some details of sampling, biomass estimation equations and periodicity of measurement of different

Table 3.3 Illustrative list of approved CDM methodologies for afforestation and reforestation projects (http://cdm.unfccc.int)

Number	Title	Above-ground biomass	Below-ground biomass	Dead wood	Litter	Soil organic carbon
AR-AM001	Reforestation of degraded land	Yes	Yes	No	No	No
AR-AM002	Reforestation of degraded lands through afforestation/reforestation	Yes	Yes	Yes	Yes	Yes
AR-AM003	Afforestation and reforestation of degraded land through tree planting, assisted and natural regeneration and control of animal grazing	Yes	Yes	No	No	No
AR-AM004	Reforestation or afforestation of land currently under agricultural use	Yes	Yes	No	No	No
AR-AM005	Afforestation and reforestation project activities implemented for industrial and/or commercial uses	Yes	Yes	No	No	No
AR-AM006	Afforestation/reforestation with trees supported by shrubs on degraded land	Yes	Yes	No	No	Yes
AR-AM007	Afforestation and reforestation of land currently under agricultural or pastoral use	Yes	Yes	Yes	Yes	No

carbon pools – this handbook supplements these methodologies so as to assist in the following aspects of carbon inventory:

- Defining the project boundary for afforestation and reforestation (A&R) project activity
- Sampling methods
- Identifying the relevant carbon pools for the selected activity
- Measuring carbon stocks under the "without-project" (baseline) scenario
- Measuring carbon stock changes under the mitigation activity scenario
- Estimating the additional net mitigation potential or net carbon stock gain
- Estimating the uncertainty in carbon inventory estimates

Although CDM is restricted only to A&R activities, it is likely that other activities such as forest conservation, forest management, grassland and agroforestry may become eligible under CDM in the future. This likelihood depends on the development of reliable and acceptable methodologies for measurement, estimation, reporting and verification of carbon stock changes from these activities. The methodologies approved up to mid-2007 along with carbon pools to be monitored are listed in Table 3.3.

3.4 Carbon Inventory for Projects Under the Global Environment Facility (GEF)

Addressing climate change concerns and, to some extent, prevention of land degradation under GEF involves increasing carbon sinks and reducing CO_2 emissions. Among a number of operational programmes, the two that are relevant to carbon inventory are *operational programme 12* on integrated ecosystem management and *operational programme 15* on sustainable land management.

Projects under these two programmes are likely to lead to conservation or enhancement of carbon sinks – through reduction or prevention of conversion of forests and grassland, felling of trees and disturbance of soil, promotion of sustainable extraction of biomass, revegetation of degraded lands, and enhancement of carbon stocks in vegetation and soils of forests, grasslands and croplands – and therefore require carbon inventory.

GEF mechanism aims at providing incremental global environmental benefits in the form of enhancement of carbon stocks in lands due to project activities. Estimation of incremental carbon benefits requires, in turn, the following estimates:

- Project area
- Carbon stock changes in the baseline, or without-project, scenario
- Carbon stock changes in the project scenario due to implementation of project activities
- Incremental carbon stock changes by deducting the carbon stocks under baseline scenario from the carbon stocks under the project scenario due to the implementation of project activities

Table 3.4 Illustrative list of approved GEF projects requiring carbon inventory

Number	Project type	Country	Carbon pools
OP 12	Programme for the Management of Forests and Adjacent Lands	Benin	Above-ground biomass and soil
	The Middle Atlas Forest Restoration Project	Morocco	Above-ground biomass and soil
OP 15	Ecosystem Restoration of Riparian Forests in Sao Paulo	Brazil	Above-ground biomass and soil
	Sustainable Land Management for Mitigating Land Degradation, Enhancing Agricultural Biodiversity and Reducing Poverty	Ghana	Above-ground biomass and soil
	Integrated Land-use Management to Combat Land Degradation and Deforestation in Madhya Pradesh	India	Above-ground biomass and soil
	Forest Protection and Reforestation	Kazakhstan	Above-ground biomass
	Programme Partnership for Sustainable Forestland Management: Phase 1	Vietnam	Above-ground biomass

Estimation of "incremental" carbon benefit, although it may appear similar to that of "additionality" under CDM, is not governed by any approved methodologies. There is no strict definition of baseline scenario for estimating the incremental carbon benefit. Broad guidelines for estimating carbon additionality are given in www.undp.org (Pearson et al. 2005a). Some illustrative projects under GEF operational programmes requiring carbon inventory are given in Table 3.4.

3.5 Carbon Inventory for Forest, Grassland and Cropland Development Programmes and Projects

Globally, the rate of net forest loss was 7.3 Mha a year during 2000–2005 (FAO 2006). According to the Millennium Ecosystem Assessment scenarios (MEA 2005), forest area during 2000–2050 is projected to increase in the industrialized world by 60–230 Mha and decrease in the developing world by 200–490 Mha. Biodiversity, particularly from tropical forests, grassland and wetland ecosystems will continue to decline. Further, many tropical regions experience shortages of fuelwood, charcoal and construction timber, particularly for use by the vast rural communities (Ravindranath and Hall 1995). Growing demand for land for crop production and commercial cattle rearing, demand for tropical timber and more recently bioenergy has led to loss of forests in many tropical countries. Land degradation affects nearly 2,000 Mha, which includes nearly 700 Mha of land subjected to overgrazing and 550 Mha of poorly managed cropland subjected to soil erosion (UNEP 2002). Thus, in response to concerns about land degradation and loss of tropical forests and biodiversity as well as about shortages of fuelwood and construction timber, a large number of forest conservation and land development programmes and projects have been formulated and implemented all over the world. The area brought under afforestation and reforestation programmes globally was 2.8 Mha annually during 2000–2005 and the total area under plantations was about 140 Mha in 2005 (FAO 2006). Apart from degradation and loss of forests, degradation of grassland and cropland and expansion of deserts are other land-related environmental concerns. Globally, a large number of programmes have been launched for conservation, reclamation and development of grassland and cropland and for halting desertification.

The main objectives of land-based conservation and development programmes in forest, grassland and cropland categories are not directly linked to addressing climate change; these programmes are aimed at conserving biodiversity, protecting watersheds, reducing pressure on forests, increasing supply of fuelwood or industrial wood and improving soil fertility to increase grass or crop productivity. However, all these programmes and projects indirectly lead to conservation or expansion of carbon pools in biomass and in soil. Improvement in biomass and soil carbon is the most important indicator of impacts of the programmes on all land-use categories, and methods and guidelines for carbon inventory are required for the following:

- Estimation of the status of land-use categories selected for project activities before implementation of the programmes
  - Above-ground biomass stock and growth rates of trees and productivity of grass or annual or perennial crops
  - Soil fertility, in particular soil organic matter and nitrogen
- Estimation of improvement due to conservation or development activities such as afforestation, forest conservation, soil and water conservation, improved grassland management, shelterbelt and agroforestry programmes, which can be measured in terms of
  - Increase in tree biomass, grass and crop productivity, particularly carbon stock in above-ground biomass and growth rates
  - Increase in soil organic matter and improvement in carbon to nitrogen ratio

A large number of forest, grassland, agroforestry and land reclamation programmes are being implemented with support from national, bilateral and multilateral funding agencies. An illustrative list of land-based development programmes requiring carbon inventory is given in Table 3.5. Such programmes require the following activities:

- Making preliminary estimation and projection of improvements in biomass production or growth rates and soil organic matter or fertility improvements at the project proposal preparation stage
- Periodic monitoring and estimation of impact of project activities on biomass production or growth rates and soil organic matter content

Table 3.5 Illustrative list of forest, grassland and cropland development programmes requiring carbon inventory

Agency	Project title or type	Country/ region	Carbon pools for inventory
Asian Development Bank	- Sustainable Agroforestry Systems for Livelihood Enhancement of the Rural Poor	Lao PDR	Above-ground biomass and soil
UK Department for International Development	- Control of Soil Acidity in Agroforestry Systems	Eastern and Middle Africa	Soil carbon
World Bank	- Pico Bonito Sustainable Forests Project	Honduras	Above-ground biomass and soil
	- Forest Sector Development Project	Vietnam	
	- Forest Sector Development Project; Supplemental Credit	Vietnam	
	- Sustainable Forestry Pilot Project	Russian Federation	
	- Maharashtra Forestry Project	India	

3.6 Conclusions

The term "carbon inventory" became popular largely under the United Nations Framework Convention on Climate Change, where carbon stock changes or CO_2 emissions and removals are estimated for land-use sectors as part of the national GHG emissions inventory process. Under the Convention, carbon inventory is also required for estimating the national and global carbon mitigation potential of land-use sectors and the mitigation achieved under land-based programmes or projects. Carbon inventory is required for programmes and projects in all land-use categories, namely forest land, grassland, cropland and settlements (see, e.g. Tables 3.1–3.3) for addressing climate change as well as for non-climate change-related programmes including the following:

- National greenhouse gas inventory programme
- Carbon mitigation programmes and projects
- Roundwood production programmes and projects
- Land reclamation programmes such as agroforestry, zero tillage, shelterbelt and watershed

Carbon inventory, in all land-use categories and under all programmes and projects, involves estimation of changes in carbon stocks in biomass and in soil and of CO_2 emissions and removals, and this book provides detailed methods and guidelines for doing so.

Chapter 4
Carbon Pools and Measurement Frequency for Carbon Inventory

Global carbon cycle involves exchange of CO_2 between the atmosphere and the biosphere, apart from oceans. Plants fix CO_2 from the atmosphere during photosynthesis to produce organic matter, which is stored in above- and below-ground parts. Bulk of the biomass in above- and below-ground plant parts is eventually transferred to the dead organic matter pool or it is oxidized or burnt. Dead organic matter, which consists of deadwood (standing as well as fallen) and litter, is either decomposed or oxidized, or stored for longer periods above or below the ground as detritus. CO_2 fixed by plants ends up in soil as organic matter or in finer forms as humus through the process of decomposition. Thus, CO_2 removed from the atmosphere is stored as dead and living biomass or soil carbon in the biosphere.

A carbon inventory, for carbon mitigation as well as forest conservation and land development programmes and for greenhouse gas inventory programmes and projects, requires estimation of stocks of carbon pools in biomass and in soil for a given period. There are five carbon pools, and measurement, monitoring and projection of changes in stocks of carbon in all the five carbon pools may be desirable. However, the cost of monitoring all the carbon pools is likely to be high. Further, stocks of some of the carbon pools may not change or change only marginally during the period selected for monitoring or projection. Therefore, the most cost-effective way of carbon inventory is to identify and monitor the key carbon pools that are likely to be impacted by the project activities or as a result of human intervention involving land-use change, conservation practices, planting trees or grasses, improved management practices, harvesting rates, cultural operations and so on.

IPCC (2003, 2006) has defined five carbon pools for greenhouse gas inventory. Even the Marrakech Accord of the United Nations Framework Convention on Climate Change (UNFCCC) has included five carbon pools, for estimating the impacts of land-use change and forestry activities:

- Living biomass
  - Above-ground
  - Below-ground (roots)

- Non-living biomass
  - Litter
  - Deadwood
- Soil organic carbon

Harvested wood products (HWP) could also be considered as a carbon pool, and some countries estimate and report stocks of HWP in their national GHG inventory. When the carbon "Stock–Difference" method is adopted, HWP pool is covered under the biomass carbon pools. Accordingly, this handbook focuses on the five carbon pools and presents their features, their importance for different land-use systems, programmes and projects and the frequency of monitoring.

4.1 Features of Carbon Pools

4.1.1 Distribution of Different Carbon Pools

The two main carbon pools are biomass and soil carbon. Biomass is defined as the total quantity of live and inert or dead organic matter, above and below the ground, expressed in tonnes of dry matter per unit area, such as a hectare. Soil carbon is carbon held in soil as organic matter, humified material and in stable structures such as charcoal. Biomass is converted to carbon by multiplying it with a carbon fraction of dry matter. The exact value of the fraction varies within a small range for different species and components of plants, and is usually about 0.5 (IPCC 2006).

Total carbon = biomass carbon + soil carbon
Biomass carbon = above-ground biomass carbon + below-ground biomass carbon + dead organic matter carbon

An illustrative distribution of different carbon pools in forests of different regions is given in Table 4.1. In Africa, for example, carbon in living biomass is predominant, accounting for about 60%, followed by soil carbon (approximately 34%), whereas in Europe, soil carbon is the predominant fraction (64%), and living biomass accounts for only 25% of the carbon. Thus, the proportions of living biomass and soil carbon vary with the region but together account for more than 90% of the total biomass. The share of deadwood and litter together is less than 11% in all regions. Carbon in the litter pool is less than 5% in all regions (Table 4.1), and soil carbon is the dominant pool in grassland and cropland systems (Table 2.2).

A study of different forest types in southern India estimated the stocks of different carbon pools (Table 4.2). Above-ground biomass accounted for 25–46% of the total carbon stocks. Below-ground biomass accounted for less than 12% of the total carbon stocks in all the forest types. Soil carbon dominated the scrub forest (68%), southern thorn forest (56%) and evergreen forest (54%). Dead organic matter carbon pool was not reported in this study. When dead organic matter was excluded, the share of above-ground biomass and soil carbon together was as much as 88–95% of the total carbon pools. In absolute terms, the total biomass and soil

Table 4.1 Distribution (%) of carbon among different pools in forests and other wooded lands (FAO 2006)

	Carbon in				
Region/subregion	Living biomass	Deadwood	Litter	Soil	Total
East and South Africa	63.5	7.5	2.1	–	73.1
North Africa	26.0	3.3	2.1	33.5	64.9
West and Central Africa	155.0	9.8	2.1	56.0	222.9
Total Africa	95.8	7.6	2.1	55.3	160.8
	(59.5)	(4.6)	(1.6)	(34.3)	(100.0)
E. Asia	37.0	5.0	–	–	41.9
South and South-east Asia	77.0	9.0	2.7	68.4	157.1
West and Central Asia	39.0	3.6	11.4	41.0	95.8
Total Asia	57.0	6.9	2.9	66.1	132.9
Total Europe	43.9	14.0	6.1	112.9	176.9
	(24.8)	(7.9)	(3.4)	(63.9)	(100.0)
Caribbean	99.7	8.8	2.2	70.5	181.2
Central America	119.4	14.4	2.1	43.3	179.2
North America	57.8	8.8	15.4	35.8	117.8
Total North and Central America	60.1	9.0	14.8	36.6	120.6
Total Oceania	55.0	7.4	9.5	101.2	173.1
Total South America	110.0	9.2	4.2	71.1	194.6
World average	71.5	9.7	6.3	73.5	161.0
	(44.4)	(6.1)	(3.9)	(45.6)	(100.0)

Figures in parentheses are percentages of the row totals

Table 4.2 Carbon stocks (tC/ha) in different pools in different forest types of Namakkal district of Tamilnadu (Ramachandran et al. 2007) and Varanasi (Misra 1972), India

	Varanasi	Tamilnadu				
Carbon pools	Deciduous	Evergreen	Deciduous	Secondary deciduous	Southern thorn	Euphorbia scrub
AGB	102.7	122	100	96	52	22
	(40.5)	(36.2)	(41.6)	(46.8)	(38.5)	(24.9)
BGB	17.1	31	25	24	6	5
	(6.7)	(9.3)	(10.7)	(12.1)	(5.0)	(6.4)
Deadwood	3.8	–	–	–	–	–
	(1.5)					
SOC (0–90 cm)	129.5	184	114	84	76	63
	(51.3)	(54.5)	(47.7)	(41.1)	(56.5)	(68.7)
Total	253.2	337	240	205	136	91
	(100.0)	(100.0)	(100.0)	(100.0)	(100.0)	(100.0)

AGB; above-ground biomass; BGB; below-ground biomass; SOC; soil organic carbon. Figures in parenthesis indicate percentages carbon stocks varied from 91 tC/ha in scrub forest to 337 tC/ha in evergreen forests (Table 4.2). Data from a deciduous forest in Varanasi, northern India, showed that dead organic matter pool is insignificant and above-ground biomass and soil carbon together account for 92% of the total carbon stocks.

4.1.2 Definition of Carbon Pools

The definitions of carbon pools according to IPCC (2006) are given in Table 4.3.

Above-ground biomass (AGB) Above-ground biomass is expressed as tonnes of biomass or carbon per hectare. Above-ground biomass is the most important and visible carbon pool, and the dominant carbon pool in forests and plantations, although not in grasslands and croplands. Above-ground biomass is given the highest importance in carbon inventory and in most mitigation projects and is the most important pool for afforestation and reforestation CDM projects under the Kyoto Protocol as well as any inventory or mitigation project related to forest lands, agroforestry and shelterbelts in croplands. The above-ground biomass is often the only carbon pool measured or estimated in roundwood production projects. The methods and models for measuring and projecting above-ground biomass are also the most

Table 4.3 Definition of carbon pools according to IPCC (2006)

Pool		Description
Living biomass	Above-ground biomass	All biomass of living vegetation, both woody and herbaceous, above the soil including stems, stumps, branches, bark, seeds and foliage
	Below-ground biomass	All biomass of live roots. Fine roots of less than 2 mm diameter (the suggested minimum) are often excluded because these often cannot be distinguished empirically from soil organic matter
Dead organic matter	Deadwood	All non-living woody biomass not contained in the litter, either standing, lying on the ground, or in the soil. Deadwood includes wood lying on the surface, dead roots, and stumps larger than or equal to 10 cm in diameter
	Litter	All non-living biomass with a size greater than the limit for soil organic matter (the suggested minimum is 2 mm) and less than the minimum diameter chosen for deadwood (e.g. 10 cm) lying dead and in various states of decomposition above or within the mineral organic soil. This includes the litter layer as usually defined in soil typologies. Live fine roots above the mineral or organic soil (of less than the suggested minimum for below-ground biomass) are included whenever they cannot be empirically distinguished from the litter.
Soil	Soil organic matter	Organic carbon in mineral soils to a specified depth chosen and applied consistently through a time series. Live and dead fine roots within the soil (of less than the suggested minimum for below-ground biomass) are included wherever they cannot be empirically distinguished from the soil organic matter.

developed compared to other carbon pools. In non-forest land-use systems such as cropland and grassland, biomass predominantly consists of non-woody perennial and annual vegetation, which makes up a much smaller part of the total carbon stock in the ecosystem than that in forest lands. The non-woody biomass is part of the annual carbon cycle and is subjected to turnover every year or every few years and hence net biomass carbon stock may remain more or less constant, although stocks may diminish over time because of land degradation.

Below-ground biomass (BGB) Below-ground or live root biomass is expressed as tonnes of biomass or carbon per hectare. Roots play an important role in the carbon cycle as they transfer considerable amounts of carbon to the ground, where it may be stored for a relatively long period of time. Although roots can extend to great depths, the greatest proportion of the total root mass is confined to the top 30 cm of the soil surface. Carbon loss and accumulation in the ground is intense in the top layer of the soil profile, which indicates that this should be the focus in sampling (Ponce-Hernandez et al. 2004). In many land-use systems such as grasslands and croplands, however, this pool may not be important. Further, below-ground biomass in grassland and cropland under annual crops is part of the annual carbon cycle, and need not be measured. Below-ground biomass is the least researched or measured carbon pool because of the difficulty in measuring or modelling of the stock or growth rates: estimating it requires uprooting of trees and grass and disturbs topsoil, which is destructive in normal circumstances, and most often, the quantity is estimated as a proportion of above-ground biomass.

Deadwood Deadwood is not a dominant carbon pool in any land-use system and usually accounts for only about 6% of the total carbon stock in forest and other wooded lands (Table 4.1). Deadwood includes naturally dead trees, both standing and fallen, and those killed as a result of pest attack, wind damage and human intervention, but excludes naturally falling woody and non-woody litter or biomass. The two fractions, namely deadwood and litter, are normally differentiated on the basis of size: a certain minimum diameter is usually stipulated as the cut-off point. Deadwood may occur largely in natural forests, but is rare in new plantations, agroforestry systems, savannah, grasslands and croplands.

Litter The layer of organic debris, dead plant material fallen or removed and plant parts not attached to plants are considered litter. Build-up of litter is a natural process in which woody and non-woody parts of trees and shrubs dry up and fall to the ground (floor of the forest or of a plantation); the process is also part of the overall process of turnover of forest biomass. Litter is not a major carbon pool because it usually accounts for only 6–8% of plant biomass (Whittaker and Likens 1973; Bazilevich 1974) and sometimes even less (Table 4.1).

Soil carbon Soil organic matter is defined as organic carbon in mineral soils to a specified depth. The generic term for all organic compounds in the soil is particles that are not living roots or animals. As dead organic matter is fragmented and decomposed, it is transformed into soil organic matter. It includes a wide variety of materials that differ greatly in their residence time in soil: some of them are easily decomposed by microbial organisms and return the carbon to the atmosphere but some of the soil organic carbon is converted into recalcitrant compounds (e.g. organic–mineral complexes) that decompose slowly and may remain in soil for decades or centuries or even longer. Fires often result in the production of small amounts of so-called black carbon, a nearly inert carbon fraction with turnover times that may span several thousand years (IPCC 2006). Within a given land-use system, such as cropland and grassland, management practices can have a significant impact on storage of soil carbon. Management practices and other forms of disturbances can alter the net balance between carbon input and carbon losses from the soil. Input to soil carbon stock can come from higher plant production. When native grassland or forest land is converted into cropland, 20–40% of original soil carbon stock can be lost (Mann 1985; Davidson and Ackerman 1993; Ogle et al. 2005). Although both organic and inorganic forms of carbon are found in soil, land use and management typically has a larger impact on organic carbon stocks, and this handbook accordingly focuses only on organic form of carbon.

4.1.3 Flux of Carbon Pools

CO_2 fixed by plants during photosynthesis is transferred across pools such as litter, deadwood and soil carbon. Carbon cycle in a land-use system includes changes in carbon stocks due to both continuous processes (growth and decay) and discrete episodic events such as disturbances (fire, harvest, land-use change and pest attack). A generalized flow chart of flux of carbon pools is given in Fig. 4.1 (IPCC 2006). All fluxes can be accounted for by estimating the stocks of all the carbon pools at

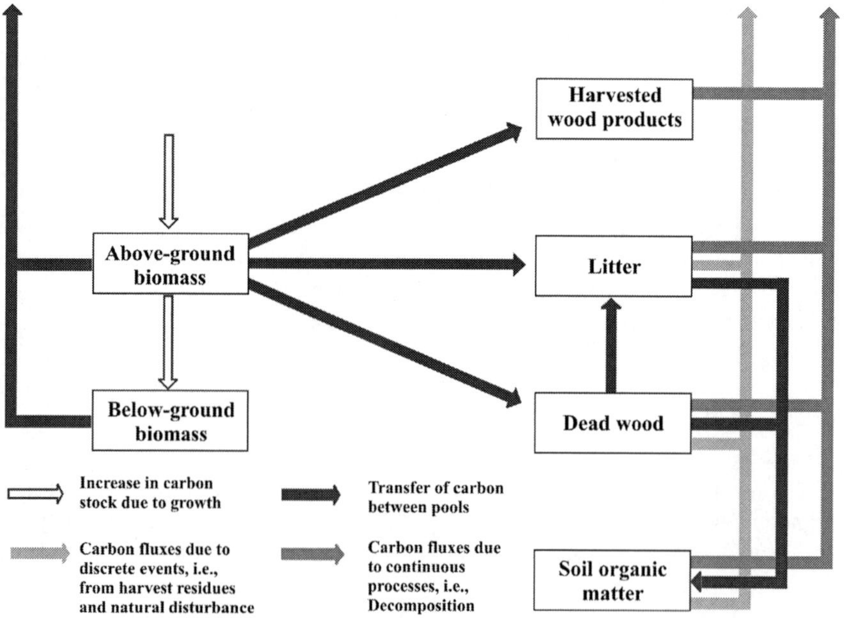

Fig. 4.1 Carbon pools and pathways of carbon flows (IPCC 2006)

two points in time. The annual carbon stock change for a given land-use system or land-use change or management practice is estimated as the sum of changes in all the pools, expressed as tonnes of carbon per hectare per year (tC/ha/year) (refer to Chapter 9 for equations).

After defining the carbon pools, and their proportions and fluxes, the next question to be addressed is this: Which among the five carbon pools should be selected for a mitigation project, a roundwood production project or a national GHG inventory process? Section 4.2 discusses the criteria for selection, and Section 4.3 assesses the importance of different carbon pools for different types of projects.

4.2 Criteria for Selection of Carbon Pools

Why select carbon pools for inventory Carbon inventory, in principle, involves estimation of changes in stocks of all the carbon pools or emissions and removals from all the pools. However, not all carbon pools are relevant to all land-use categories, or project types, and the general practice is to estimate the changes in the stock of a key pool or a set of key pools.

Estimation of changes in stocks of all the carbon pools is expensive – the goal should be to maximize the cost-effectiveness of a carbon inventory, which is to obtain the highest possible accuracy for a given level of resource or human effort; the latter is always a limiting factor. IPCC (2003, 2006) provides an approach called "key category analysis" to select the key pools that are critical for a greenhouse gas inventory (see Chapter 16). The choice of a carbon pool or pools for monitoring or estimation for different land-based programmes and projects depends on the land-use system, goals of the project, activities implemented and the period selected for monitoring. A broad approach to selection of carbon pools for an inventory may involve the following factors.

(i) *Land-use system* The proportion of stocks of different carbon pools in a unit area of land varies with the land-use system: above-ground biomass is likely to be the most impacted carbon pool in land-use systems such as forests, plantations and agroforestry whereas soil carbon pool is likely to be impacted most by activities in cropland, grassland and savannah projects. Similarly, litter or deadwood pool is likely to be nearly absent in cropland or grassland systems and may be important only for natural forests.
(ii) *Project goals and activities* The importance of different carbon pools will vary with the project goals of interest to project managers. Carbon mitigation projects directly aim at enhancing carbon stocks or reducing the emission of CO_2 from a unit area, and it is important to consider all the potential pools likely to be impacted by project activities that contribute to the aim. In a land reclamation project, the main goal may be to enhance soil organic matter or carbon density of the area to improve soil fertility, making it the only carbon pool of significance. In a community fuelwood plantation or commercial timber project, the main objective is to enhance roundwood production, making above-ground biomass the key pool. Tree planting, promotion of natural regeneration,

protection, halting felling of trees, soil conservation and changing harvesting or grazing practice are all examples of project activities that may determine the carbon pool most affected by them. Project activities are of course determined by the goals of the project. For example, both halting the felling of trees and protection and promotion of natural regeneration will impact above-ground biomass pool the most whereas improved grazing or soil conservation practices will impact soil carbon pool the most.

(iii) *Period considered for monitoring* The rate of accumulation of carbon varies among different pools even for a given project-type at a given location. In an afforestation and reforestation project, above-ground biomass builds up faster, on an annual basis, than any other carbon pool. Soil organic carbon increases slowly. Therefore, the period of monitoring selected for estimating carbon stock gains in a project is also important in determining the carbon pools to be selected for monitoring. In short-term projects, of say 2–3 or 5 years – e.g. an afforestation, a reforestation or an agroforestry project – soil carbon, litter and deadwood pools are unlikely to be key carbon pools. Normally, soil carbon pool requires monitoring only over longer periods.

(iv) *Cost of monitoring* In theory and in practice, all carbon pools can be measured annually using standard methods. Periodical and, in particular, annual monitoring of all the pools in any land-based project may have large cost implications for the whole project. Cost is likely to be the overriding criterion not only in selection of the carbon pools, but also in deciding on the methods and frequency of monitoring. The cost of monitoring different carbon pools is different: for example, below-ground biomass is more expensive to monitor than above-ground biomass.

A project manager has to balance various factors such as the cost and the accuracy required with respect to main goals of the project, activities involved and their likely impacts on different carbon pools. Further, decisions on key pools and the number of pools to be monitored for carbon mitigation projects will be different from those for other land-based conservation and development projects.

4.3 Key Carbon Pools for Different Programmes and Projects

The key carbon pools to be monitored will be primarily determined by the main goal or objective of the project, and potential key carbon pools for different carbon mitigation, land conservation and development, and roundwood production programmes and projects are considered in this section.

4.3.1 Carbon Mitigation Projects

The main goal of carbon mitigation projects is to maximize gains in carbon stocks and minimize emissions of CO_2 from land-use systems. For example, in avoided deforestation projects, the main goal is to reduce the emissions of CO_2 from forest land conversion or harvest of trees, whereas in afforestation and reforestation projects, maximizing the biomass and soil carbon stocks is the main focus. The selection of carbon pools depends on the land-use system considered for mitigation. The key carbon pools for different types of projects are given in Table 4.4.

(i) *Avoided deforestation* Avoided deforestation is a dominant carbon mitigation opportunity in land-use sectors (IPCC 2007c) and all the five carbon pools are likely to be important, necessitating monitoring of all of them. Further, the selection of carbon pools may be influenced by the following considerations:

- Avoided deforestation leading to halting land-use change, from forest land to cropland for example, would impact all the carbon pools in the project area since all the biomass pools may be removed and burnt. Soil carbon will be lost due to land preparation or disturbance to topsoil, particularly if forest land is converted to cropland.
- Deforestation leading to degradation of forest land with no land-use change or land conversion, may require monitoring of only above-ground biomass if commercial harvesting is practiced, leading to removal of all standing trees. Litter and deadwood may have to be monitored if these pools are removed or burnt. Soil carbon is unlikely to be impacted if land-use remains unchanged or topsoil remains undisturbed.

(ii) *Afforestation and reforestation, including bioenergy plantations* The key carbon pools likely to be impacted because of afforestation and reforestation are above-ground biomass, below-ground biomass and soil carbon. The accumulation of litter and deadwood pools is unlikely to be significant in the short term. If forest land is converted to plantations, soil carbon is likely to be impacted significantly because of land preparation practices. Any disturbance to soil as

Table 4.4 Key carbon pools for carbon mitigation and other land-based projects

		Carbon pool				
Project category	Project type	Above-ground biomass	Below-ground biomass	Litter	Deadwood	Soil
Carbon mitigation	Avoided deforestation	***	***	***	***	***
	Afforestation and reforestation	***	**	*	*	***
	Bioenergy plantations	***	***	*	-	***
	Forest management for timber	***	*	*	*	*
	Grassland management	*	*	-	-	***
Round wood pro-duction	Commercial plantations	***	*	-	-	***
	Community forestry projects	***	-	-	-	**
Other land-based	Agroforestry	***	*	-	-	*
	Shelterbelts	***	*	-	-	*
	Watershed	**	-	-	-	***
	Land reclamation	*	*	-	-	***

*** = high impact, ** = medium impact, * = low impact, - = no impact or marginal impact part of land preparation will have significant implications for the soil carbon pool. However, if reforestation is carried out through protection and promotion of natural regeneration, without disturbing the topsoil, soil carbon pool is unlikely to be impacted in the short term, say less than five years.

(iii) *Forest management* Activities under forest management may include sustainable logging, fertilizer application and thinning. Forest management does not involve any land-use change or disturbance to topsoil. Thus, the only carbon pool likely to be impacted significantly is above-ground biomass in the short term and soil carbon in the long term.

(iv) *Grassland management and land reclamation. Above-ground biomass, litter and* deadwood are unlikely to be impacted by project activities leading to improved grassland or annual cropland or to better management of degraded land. Even under the baseline situation, tree biomass may not be present in any significant quantity in croplands, grasslands or savannahs. The main goal is to improve soil fertility and grass or crop production. Above-ground biomass and root biomass in grasslands is an important carbon pool but it is part of the annual cycle, and soil carbon is likely to be the only key pool.

The carbon pools likely to be impacted by a given type of mitigation project may sometimes depend on the location, dominant species and management practice. Thus, the project manager has to make an expert judgment on which pools to be monitored based on location-specific factors as well as the general guidance provided in this handbook.

4.3.2 Roundwood Production, Land Conservation and Development Projects

Programmes and projects for roundwood production and for land conservation and development are aimed neither at reducing CO_2 emissions nor at increasing carbon stocks but seek to enhance or conserve forest biodiversity, improve soil fertility and increase roundwood production. However, all such projects impact carbon pools, and monitoring or evaluation of these projects actually requires estimation of changes in some – though not all – of the carbon pools. The focus is largely on enhancing production of above-ground biomass or improvement of soil fertility.

Roundwood production programmes or projects The main goal of programmes or projects that seek to produce roundwood is to produce fuelwood, industrial roundwood or sawn timber. Above-ground tree biomass is harvested periodically according to a rotation cycle, although soil is unlikely to be disturbed if coppicing plants such as eucalyptus and teak are planted. Therefore, above-ground biomass is the key pool of interest to the project manager, and soil carbon may also be of interest but more as an indicator of soil fertility. If the periodical and cyclical harvesting is sustainable, significant biomass is likely to be present at any time in the plantations, requiring periodic monitoring of the standing stock. Thus, above-ground biomass and, in some cases, soil carbon are the key pools for roundwood production programmes.

Land reclamation projects The main goal of projects aimed at land reclamation is to improve soil fertility, increase crop and grass production and even improve livelihoods – carbon mitigation is not the main focus, although mitigation occurs in all land conservation and development projects. The project managers are likely to focus on improving soil quality. However, monitoring soil fertility requires monitoring of soil organic carbon pool.

Illustrative lists of key carbon pools to be monitored for different project types are given in Table 4.4. The following features are apparent:

- Soil carbon is a key pool for all projects, since land or soil fertility improvement and conservation are important components of all land development projects
- Biomass pool, particularly above-ground biomass, may be impacted by agroforestry and shelterbelt programmes

4.3.3 National Greenhouse Gas Inventories

A national GHG inventory involves estimation of CO_2 emissions and removals from different land-use categories for the selected inventory year. IPCC (2003, 2006) guidelines cover estimations from all the five carbon pools, according to different land-use categories. Further, the guidelines also include the "key category analysis", which can be used for identifying the key carbon pools for the inventory (Chapter 16), although the choice may vary with the country, region and land-use category. The key category analysis helps in focusing resources and efforts on the dominant carbon pool or pools.

4.4 Frequency of Monitoring of Carbon Pools

The rate of accumulation of different carbon pools varies with project types and activity. The suggested frequency for monitoring different pools is given in Table 4.5. The list is only illustrative, and project managers have to use expert judgment to decide on the frequency, depending on the project objectives, vegetation types, dominant species, silvicultural practices and local conditions. Features of different carbon pools that determine the frequency of monitoring are presented in this section.

4.4.1 Above-Ground Biomass

Above-ground biomass, the most important carbon pool for all land-use systems and projects involving trees, is likely to change frequently, even annually, much faster than other pools for all projects involving tree planting. Such a change requires frequent monitoring of the pool. It can be observed from Table 4.5 that all forest- and plantation-related projects require monitoring of above-ground biomass

Table 4.5 Frequency of monitoring of carbon pools in carbon mitigation and other land-based projects

Project category	Project type	Monitoring frequency of carbon pools (years)				
		Above-ground biomass	Below-ground biomass[a]	Litter	Dead wood	Soil
Carbon mitigation	Avoided deforestation	Annual	–	3–5	3–5	Annual initially
	Afforestation and reforestation	Annual	–	3–5	3–5	3–5
	Bioenergy plantations	Annual	5	3–5	–	3–5
	Forest management	2–3	–	3–5	3–5	5
	Grassland management	2–3	–	–	–	3
Roundwood production	Commercial plantations	Annual	–	3–5	–	5
	Community forestry	Annual	–	3–5	–	5
Other land-based	Agroforestry	Annual	–	–	–	5
	Shelterbelts	Annual	–	–	–	5
	Watershed	Annual	–	–	–	2–3
	Land reclamation	2–3	–	–	–	2–3
	Grassland development	2–3	–	–	–	2–3

[a]Below-ground biomass measurements may be feasible only if uprooting of sample trees or grasses is feasible. This pool is estimated based on the above-ground biomass values for tree-based projects.

either annually or once in 2–3 years. Intensively managed bioenergy plantations or commercial plantations also require annual monitoring of above-ground biomass pool due to the likely high growth rates and short rotation of the species planted. Further, frequency depends on the stock in above-ground biomass pool at the pre-project implementation phase. If the stock is zero or insignificant, as with most afforestation and reforestation or bioenergy plantations, its annual accumulation is likely to be significant and should be measured annually whereas in the case of forest management projects, above-ground biomass is likely to be present in substantial quantities at project initiation (baseline scenario), and additions will amount to only a small fraction of the initial quantities – annual monitoring, therefore, is not required. Thus, the frequency with which above-ground biomass stock needs to be monitored depends on the stock at the beginning of the project, the likely rate of growth of biomass (which depends on the species and silvicultural practices) and the objective of the project and normally ranges from yearly to once in 2–3 years.

4.4.2 Below-Ground Biomass

Below-ground biomass is unlikely to be measured for majority of the forest and tree plantation projects because of the complexity of the methods and high cost (see Chapter 11) – it is more usual to calculate the quantity of below-ground biomass by taking it as a proportion or function of the above-ground biomass pool. When the quantity can be measured, say for afforestation projects, the frequency could be once in five years. Forest management activities are unlikely to impact below-ground biomass, since root biomass is not disturbed and no fresh planting undertaken. With grassland and land reclamation projects, root biomass is likely to be part of the annual carbon cycle. Thus, in all projects involving tree planting, below-ground biomass could be estimated as a proportion of above-ground biomass, using data on root/shoot ratios or a biomass equation, and the frequency of estimation will be the same as that of the above-ground biomass.

4.4.3 Litter and Deadwood Biomass

Biomass in litter and deadwood contributes only a small fraction, usually less than 10%, of the total carbon stock of forests and plantations. The frequency of estimating these pools depends on whether the two pools are left undisturbed (as in forests and plantations) or removed periodically as part of cultural operations or for fuel-wood purpose. If woody litter or deadwood is removed, it is suggested that it be weighed whenever it is removed: if left undisturbed, the stock could be estimated every three to five years. Carbon mitigation projects may require estimation of carbon stocks in these pools at two points in time to calculate changes in the stock. If so, the frequency could be the same as that for estimating above-ground biomass since the additional effort required is marginal. For non-tree-based projects such as grass-land development or land reclamation, these two pools need not be measured at all; even for agroforestry and shelterbelt projects involving tree planting, the incremental stock change in these two pools is likely to be too insignificant to merit measurements.

4.4.4 Soil Carbon

As mentioned earlier (Section 4.1.1), soil carbon is next only to above-ground biomass in terms of its contribution to the incremental total carbon stock in forests and plantations; in grassland or cropland projects, soil carbon is the predominant carbon pool. Soil carbon is likely to accumulate in forest and plantation projects, and the annual incremental stock over the baseline stock is likely to be very small, making it difficult to measure. For example, in degraded lands considered for afforestation and reforestation, the stock of carbon in soil at the time of planting could be 30–60 t/ha, and annual incremental addition due to afforestation is likely to be low at 0.25–1 t/ha. The difficulty in measuring such a small addition to a base stock is compounded by errors and uncertainties in measurement methods – the small estimated value could well be within the range of error. The same rationale applies to grassland or land reclamation projects. In avoided deforestation and forest management projects, soil carbon stock is unlikely to change measurably within 2–5 years because such projects do not involve change in land use or disturbance to topsoil.

However, for avoided deforestation or grassland conservation projects involving change in land use, such as conversion of forest land or grassland to cropland, soil carbon stock will be significantly impacted, requiring more frequent monitoring, at least in the initial years of the change in land use. In general, for most projects, soil carbon pool can be measured once in, say, five years.

4.5 Conclusions

Carbon inventory is required for all land-based programmes and projects. Five carbon pools have been identified for such projects, particularly for carbon mitigation projects and mechanisms as well as for a national inventory of GHGs. The ideal approach will be to measure and monitor all the carbon pools every year. However, in practice, such an approach may turn out to be neither essential – all pools may not be impacted – nor cost-effective. Thus, there is a need to adopt appropriate criteria to decide which carbon pools should be monitored and how frequently. Such criteria include the following:

- System of land use such as degraded forest land, natural forest, grassland and cropland
- Type of project including carbon mitigation, afforestation and reforestation under CDM, commercial industrial roundwood production and community forestry programmes
- Goal of the project, e.g. carbon mitigation, industrial roundwood production, fuelwood production, grassland reclamation and livelihood improvement
- Pool that will be impacted (all pools or only above-ground biomass or only soil carbon or a combination of pools) and the extent of impact
- Rate of change in the pool, either slow (soil carbon) or fast (above-ground biomass)

Ultimately, it will be for project developers and managers to make expert judgments on selection of the carbon pools and frequency of monitoring by using the criteria mentioned in this handbook and also taking into account the local conditions. The goal should be to capture the maximum impact on the stocks of carbon pools cost-effectively.

Chapter 5
Carbon Inventory in Project Development, Implementation and Monitoring Phases

A carbon inventory is required for projects involving carbon mitigation, traditional forestry and roundwood production as well as for other land-based development (non-carbon) projects. This chapter focuses on the carbon inventory process at different phases of a project cycle. A typical project cycle comprises conceptualization, consultation, proposal preparation, appraisal, approval, implementation, monitoring and evaluation. Project proposals are written to seek support for investment capital or grant, technology transfer, capacity development or for conducting research. Most land-based projects involve appraisal, monitoring and evaluation of environmental, financial, social, institutional and legal aspects at multiple stages of the project cycle. The focus of this handbook is on environmental aspects, particularly assessment of carbon emissions, removals or carbon stock changes due to project activities that require a carbon inventory. Even projects not directly related to carbon mitigation require assessment of production of roundwood or grass, rates of growth of organic matter in soil and stock changes. Such agencies as the World Bank, Asian Development Bank, UN organizations, Clean Development Mechanism (CDM) Executive Board, bilateral funding agencies and national institutions have their own guidelines for developing project proposals, appraisal, monitoring and evaluation. However, the basic components of any project cycle involving carbon mitigation, roundwood production or land reclamation would involve multiple stages as illustrated in Fig. 5.1.

Estimation, projection and modelling of roundwood or carbon sequestration rates and stock changes are probably required at all stages of a project cycle. Thus, carbon inventory guidelines are required for all stages. Broadly, carbon inventory methods can be considered at two phases, namely

- *Project development or pre-project implementation phase*
    - Conceptualization
    - Proposal development
    - Review and approval
- *Project monitoring or post-project implementation phase*
    - Implementation
    - Monitoring
    - Evaluation midway through and at the end

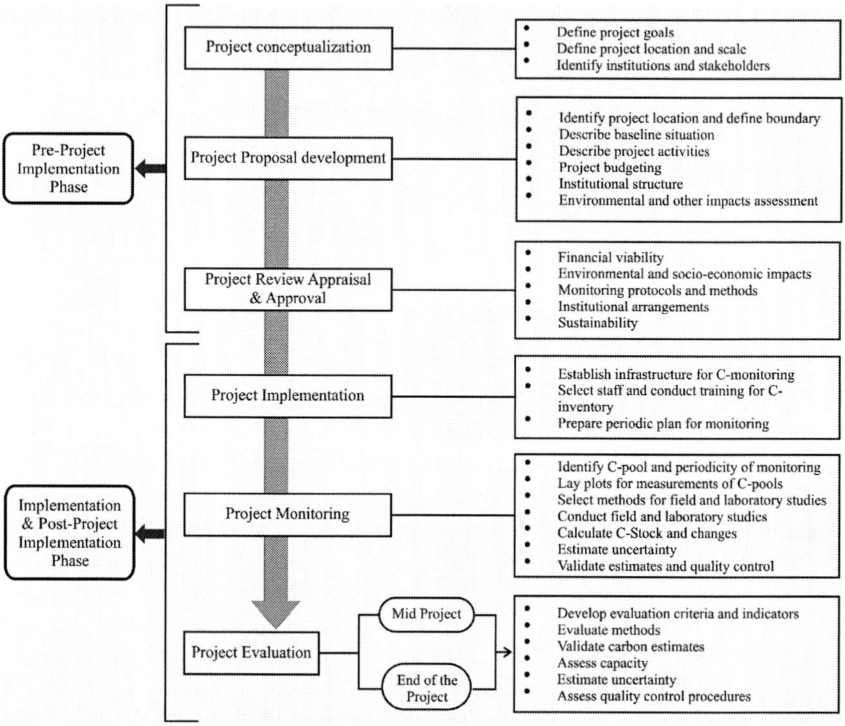

Fig. 5.1 Project cycle and activities relevant to carbon inventory

The broader guidelines required for carbon inventory projects are presented in this chapter for the pre- and post-project implementation stages. The steps described in the project cycle focus on implications for carbon inventory.

5.1 Project Conceptualization

The project conceptualization phase involves conceiving the project idea, setting goals, assessing potential locations, deciding on the scale of the project, and involvement of institutions and stakeholders, as well as consideration of the likely environmental, social and economic impacts.

Defining project goals The project could have multiple interrelated goals or a single goal. Examples of project goals are given below:

- *Environmental goals*: climate change mitigation, forest and biodiversity conservation, land reclamation and sustainable land (or forest or ecosystem) management

- *Economic goals*: production of timber, industrial roundwood, fuelwood and biomass feedstock for energy generation
- *Social goals*: improving the livelihoods of forest- and livestock-dependent communities, participatory management of natural resources including forests or grassland

The project goals are defined by involving all the concerned stakeholders. A commercial project for industrial roundwood production may have an overriding objective of maximizing harvestable roundwood, while a community forestry project is likely to have multiple goals such as biodiversity conservation, fuelwood production, land reclamation and employment or livelihood generation. The objectives of a project determine the importance of carbon inventory and the different carbon pools. A carbon mitigation project may require all carbon pools to be estimated, while that meant for producing industrial wood may focus only on above-ground merchantable biomass pool.

Location and scale of the project It is important to identify potential locations early in the project cycle for all land-based projects. The project area could be a single parcel of land or have multiple locations, such as villages, forest patches or farms. Single or multiple locations have implications for sampling. Multiple and dispersed locations of project area may require multistage stratification and a larger sample than a single parcel of the project area, even when the total project area is the same. The heterogeneity of land plays an important role in sampling and thus has implications for carbon stocks and inventory methods. The project could be a small-scale project covering a village or a few hundred hectares or it could be a large-scale project involving tens or hundreds of thousands of hectares. The scale too has implications for carbon inventory methods: a small-scale project may adopt measurement of trees using the plot method, while a large project may add remote sensing techniques for estimating biomass in addition to the plot method.

Institutions and stakeholders Goals, location and scale of a project may often depend on the stakeholders involved. In a community forestry project, the key stakeholder is the local community; a project for producing industrial roundwood may involve a single industry controlling all the resource or the project could be a joint venture of large number of farmers and an industry. The type of institutions and stakeholders has implications for carbon inventory methods. A community forestry project may adopt simple and cost-effective participatory techniques for estimating biomass (e.g. fuelwood) or carbon stocks, while a large commercial project for producing industrial roundwood may adopt remote sensing techniques for estimating roundwood production or carbon stocks.

5.2 Project Proposal Development Phase

Developing a project proposal involves clearly defining the specific objectives, describing project activities, defining the project location and area, choosing the methods for estimating and monitoring environmental and other impacts and project costs, specifying the institutional structure for the project, and selecting carbon pools and frequency for monitoring.

Identification of project location and defining scale of the project The exact location of the project, including the administrative and tenurial (e.g. user rights) boundaries, number of parcels, total area to be covered under each parcel and the entire project need to be described in the proposal. A map of the project location depicting the various land-use systems and the project sites along with geographic coordinates is required (refer to Chapter 8 for details on defining the boundary of a project).

Description of baseline scenario It is necessary to describe the land-use characteristics such as vegetation status, carbon stocks and grazing and harvest practices in the land area proposed for the project and the historical as well as the likely changes in the future in land-use and their implications for carbon stocks in the absence of the proposed project activities. It is very important to describe the methods adopted for estimating and projecting carbon stock changes under the baseline scenario (see Chapter 7 for a more detailed description of baseline scenario development).

Description of project activities An adequate description of project activities is the most critical component of the project proposal development phase and involves presenting all the activities proposed for achieving the project goals as well as for monitoring and evaluating project impacts, including carbon stock changes. Further, the project proposal should describe the phasing of different activities such as the extent of land to be planted annually, silvicultural practices and proposed harvest schedules. Some of the project activities, which have implications for carbon stock changes as well as for methods of estimating such changes, are presented in Table 5.1.

(i) *Physical activities* include fencing, boundary demarcation, locating irrigation source, arrangements for water distribution and establishment of laboratory facility for estimation of soil carbon or dry matter. Fencing could prevent livestock grazing with implications for regeneration and growth of seedlings. These activities need to be included in the project description.

(ii) *Cultural practices and management activities* could include selection of planting material, planting, fertilizer application, thinning, harvesting and transportation. The cultural and management practices have implications for carbon stocks and inventory. For example, fertilizer application could enhance biomass growth rates and land preparation could lead to loss of soil organic carbon.

(iii) *Capacity and institutional development* aspects including selection of staff for project implementation, administration and monitoring should be included in the project. Activities for building technical capacity to undertake various components of the project, including monitoring of carbon stock changes, need to be described.

(iv) *Monitoring procedures and activities* could include description of the methods to be adopted for assessment of environmental, economic and social impacts of the project activities. In addition, the methods to be adopted for periodic monitoring of carbon stock changes should also be described.

Project budget All the investment, operational, maintenance and monitoring costs on an annual basis should be estimated and presented. The project should include

Table 5.1 Implications of project activities for carbon inventory methods for broad land-use based projects

Broad project type	Activities or practices	Implications for carbon inventory methods
Avoided deforestation	Enhance crop productivity to reduce conversion of forest to cropland	- Forest and cropland area and carbon stock changes need to be monitored
	Adopt sustainable harvest practices	- Harvest levels and changes in carbon stocks in above-ground biomass need to be monitored
Afforestation and reforestation	Protection, natural regeneration	- Tree measurement and biomass estimation methods will vary for plantations of monoculture for naturally regenerated forests
	Monoculture plantation	- Developing allometric equations for biomass is feasible for short-rotation species using harvest method
		- Monitoring of carbon pools needs to be more frequent for short rotations
	Fertilizer application	- Growth rates of biomass and soil carbon need to be monitored
Bio-energy plantations	Intensive cultivation practices; land preparation, high density planting and fertilizer application	- Carbon pools need to be measured more frequently
		- Measurement of harvested biomass is feasible
Agroforestry	Plant trees in rows	- Rate of change in soil carbon likely to be small, thus monitoring can be at longer intervals
	Plant fruit orchards with low tree density	- Tree felling not feasible
Shelterbelt	Plant rows of trees to halt soil erosion	- Carbon pools in litter and deadwood may not be significant
Grassland reclamation	Improved grazing	- Monitoring soil carbon pool will be very important
	Soil and water conservation	- Aboveground biomass pool not critical

the cost of monitoring the project benefits and impacts. The monitoring cost, particularly for carbon monitoring, includes:

- Staff and training
- Field equipment and instruments for measurement
- Laboratory instruments and chemicals
- Travel
- Procurement and interpretation of data from remote sensing

Institutional structure The institutional arrangements required for implementation, management, monitoring and evaluation of project activities need to be assessed and presented in the project proposal. The institutional arrangements, roles and responsibilities of different institutions along with training and capacity development needs for carbon inventory should be incorporated in the proposal.

Assessment of environmental and other impacts The proposal should include the methods, the institutional arrangements and infrastructure needed for assessing the performance and impacts of the proposed project activities against the goals and objectives defined. The impacts to be assessed could include the following:

- *Environmental impacts such as* carbon stock changes, soil quality including organic matter status, biodiversity, vegetation regeneration and plant productivity
- *Social and economic impacts such as* timber and fuelwood production, employment and income generation, costs and benefits
- *Sustainability* of environmental and socio-economic benefits as well as the institutional arrangements

5.3 Project Review, Appraisal and Approval Phase

Procedures for project review and appraisal vary with different agencies, and the project review phase is critical in getting a project approved and financially supported. The procedures are usually complex for land-based projects due to environmental and socio-economic linkages and trade-offs. Appraisal of the project could include consideration of long-term impacts on carbon stocks, biodiversity, soil quality and water supply. The review and appraisal process could potentially focus on the following aspects:

(i) *Financial viability* of the project, as indicated by the cost–benefit analysis as well as the internal rate of return, is critical in project approval. Carbon stock gains contribute to financial viability of carbon mitigation and roundwood biomass production projects.

(ii) *Environmental impacts* form the critical basis for achieving project goals. Carbon gains would be part of most land-use based projects involving:

   - Roundwood and grass biomass production
   - Biodiversity
   - Soil quality (e.g. increasing organic matter or erosion control)
   - Carbon stock changes in different pools

(iii) *Monitoring protocols and methods* are evaluated for their reliability, accuracy, verifiability and cost-effectiveness. Estimates of carbon stock changes in land-use projects are characterized by high uncertainty. High transaction costs to reduce uncertainty of carbon inventory estimates are a limiting factor for projects under Clean Development Mechanism and Global Environment Facility. Thus, selection of carbon monitoring protocols and methods involves a trade-off between accuracy and costs, requiring an optimum balance between the two.

(iv) *Institutional arrangements* proposed in the project description for project implementation, management and monitoring will be evaluated for adequacy, efficacy and cost-effectiveness.

(v) *Sustainability* of environmental, economic and social benefits will determine whether the project is approved. Sustaining the carbon benefits will be critical for projects related to mitigating climate change.

5.4 Project Implementation Phase

All the activities incorporated in the approved project proposal and aimed at achieving the goals of the project will have to be implemented. The activities could include land preparation, planting trees or grasses, regulating grazing, protection, fertilizer application, thinning and harvesting, measuring carbon pools, capacity building and infrastructural development. Some of the activities relevant to carbon inventory include:

(i) Establishing infrastructure for carbon inventory such as laboratory facilities and permanent field plots
(ii) Selection of staff and training for carbon inventory
(iii) Procurement and interpretation of data from remote sensing
(iv) Arrangements for periodic monitoring of carbon pools
(v) Estimation of baseline carbon stocks for different land-use categories

5.5 Project Monitoring Phase

Periodic monitoring of project performance and impacts is crucial to achieving the goals of a project. Monitoring would include periodic assessment of environmental, economic and social indicators selected for the project along with project costs and benefits. The key aspects of monitoring relevant to carbon mitigation and sustainable biomass (timber and fuelwood) production and land management would include the following activities:

(i) Identification of carbon pools and periodicity of monitoring for baseline and project scenarios
(ii) Laying of plots for measurement of carbon pools under baseline and project scenarios
(iii) Measurement of parameters related to identified carbon pools and indicators in the field and laboratory for baseline and project scenarios
(iv) Calculation of carbon stocks and changes for different land-use categories and in selected carbon pools under baseline and project scenarios

(v) Estimation of uncertainty or error involved in carbon stock estimation
(vi) Validation and quality control

5.6 Project Evaluation Phase

All projects require evaluation of the achievement of project goals. This includes the impacts at the end of the project and very often at different phases of project implementation. Project evaluation is normally carried out by an outside agency not involved in project formulation and implementation. Project evaluation activities include the following:

- Developing evaluation criteria and indicators, based on the agency (GEF, UNFCCC, multilateral banks, bilateral agencies and national government ministries) as well as the project type (CDM, industrial roundwood production, community forestry, forest and biodiversity conservation and land reclamation), for assessing;
  - Carbon stock gain and roundwood production
  - Socio-economic impacts
  - Sustainability of project benefits, including carbon stock gain
- Evaluating selection of carbon pools, periodicity of measurements, methods, sampling and calculation procedures for baseline as well as project scenarios
- Validating carbon stock estimates under baseline and project scenarios
- Assessing institutional and technical capacity for monitoring carbon benefits
- Estimating the uncertainty involved and assessing quality control procedures adopted

5.7 Carbon Mitigation and Non-Carbon Land Development Projects: Implications for Carbon Inventory During Project Cycle

Carbon inventory guidelines and methods could vary for different types of projects and at various phases of a project cycle. The focus of carbon mitigation projects is different from non-carbon forest, land conservation and development projects. Thus, the importance given for carbon inventory could vary with different phases of the project cycle. However, it is important to note that carbon inventory is required generally for all land-based projects. Some of the key differences between carbon mitigation projects and other land-based projects are highlighted in Table 5.2.

Carbon inventory methods will form the primary criterion for the approval phase of carbon mitigation projects. Further, monitoring and evaluation of carbon mitigation projects will be largely focused on the reliability, accuracy and verifiability of carbon stock estimates and changes.

Table 5.2 Key differences relevant to carbon inventory at different project phases for carbon mitigation projects and non-carbon land-based projects

Project phase	Carbon mitigation projects	Forest, grassland, cropland conservation and development projects (non-carbon)
Conceptualization	Primary focus: carbon mitigation and carbon credits – global environmental benefit	Primary focus on forest and biodiversity conservation, watershed protection, commercial roundwood production
	Secondary focus: soil and biodiversity conservation	Co-benefits: carbon mitigation often not mentioned in proposal
Proposal development	Clear historical records of the past vegetation and soil carbon status needed	Historical vegetation status not so critical to project eligibility
	Project boundary impacted by project activities needs clear definition	Project boundary needed for estimating environmental and socio-economic benefits restricted to project area
	Estimation of baseline carbon stocks is critical	Baseline economic benefits, soil fertility and biodiversity critical
	Intensive plan needed for monitoring carbon stock changes	Well-defined plan needed for monitoring roundwood production and local environmental and socio-economic impacts
Project review and appraisal	Baseline and project scenario carbon monitoring methods and arrangements very critical	Monitoring plan for local environmental and socio-economic benefits critical
Implementation	Activities aimed at maximizing carbon benefits, followed by other co-benefits	Activities aimed at maximizing biomass production, biodiversity conservation and livelihood improvement
Monitoring and evaluation	Approved methodologies	Project-specific methodology: no global standards
	Additionality of carbon stock gain critical	Additionality of local environmental and socio-economic benefits critical
	All the five carbon pools need to be considered	-Above-ground biomass pools critical for roundwood production
		-Soil carbon critical for land development projects
	Large transaction cost likely for carbon inventory and monitoring	Moderate transaction cost for monitoring

5.8 Conclusions

Carbon inventory is required for most land-based projects, whether aimed at carbon mitigation or at development of forest, grassland and agroforestry and at commercial roundwood production. The focus and intensity of carbon inventory will vary based on the mechanism or programme for which a project is designed and the primary goal of the project. Carbon inventory is required at all phases of a project cycle from project conceptualization to evaluation at the end of the project. Given the high uncertainty in the estimates of carbon stock changes, the transaction costs for improving the accuracy and reliability of estimates are likely to be high. Advanced planning is required for identifying the infrastructural, staff and capacity building needs for carbon inventory process. Adequate planning and incorporation of carbon inventory monitoring methods will facilitate the approval process of a project as well as monitoring and verification of carbon benefits from a project.

Chapter 6
Methodological Issues in Land-Based Projects

All land-based projects require carbon inventory. Carbon inventory for land use, land-use change and forestry (LULUCF) projects for climate change mitigation is contentious because of both methodological issues and uncertainty in the data required to estimate gains in carbon stocks. LULUCF projects include carbon mitigation activities in three categories, namely forestry, cropland and grassland. The complexity of methods for estimation and projection of carbon stock changes leads to several methodological issues, which are important to consider at different phases of a project cycle. Examples of methodological issues are non-permanence, leakage and additionality of carbon gains. Some of these issues are also relevant to non-climate mitigation, land-based conservation and development projects. Issues such as non-permanence are unique to land-based mitigation projects. The following methodological issues are addressed in this chapter:

- Baseline
- Additionality or incrementality
- Permanence
- Leakage
- Project boundary
- Scale of projects

6.1 Baseline

All land-based carbon mitigation projects require estimation of net carbon stock gains resulting from the implementation of project activities. It is important to recognize that even in the absence of a proposed project, carbon stock will change because of natural factors or human intervention. This fact requires estimation of carbon stock changes that would have occurred in the absence of the proposed project, a situation or scenario referred to as the "baseline". A baseline needs to be developed for all projects against which project results can be compared and additional benefits estimated. According to the UNFCCC "the baseline for project activity is the scenario that reasonably represents anthropogenic emissions by sources and removal by sinks that would occur in the absence of the proposed project activity" (UNFCCC 2002). The baseline scenario is also often referred to as the reference or business-as-usual scenario. Estimating and projecting baseline carbon stocks is a difficult task since all potential future scenarios in the absence of the project must be evaluated. Establishing the baseline scenario therefore requires knowledge of the history of the given area, local socio-economic situation, and ongoing and wider economic trends (national, regional or even global) that may affect future land use and carbon stocks (Fig. 6.1).

Carbon stocks on the land when the project starts, often referred to as the base year, can serve as the baseline if it can be demonstrated that changes to business-as-usual activities in the foreseeable future (i.e. 10–30 years) are unlikely to impact carbon stocks. The baseline is established by projecting these past trends, current situation and future plans. Consequently, a baseline scenario is necessarily based on a range of assumptions.

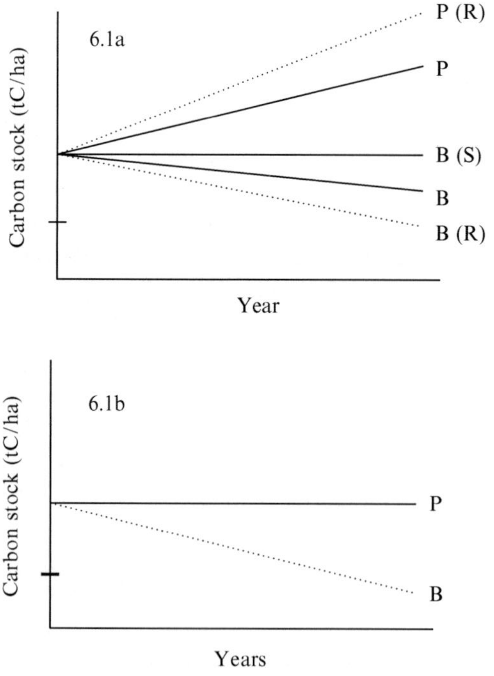

P	Project scenario carbon stocks
B	Baseline scenario carbon stocks
B (R)	Baseline scenario re-estimated carbon stocks
B (S)	Stable baseline carbon stocks
P (R)	Project scenario re-estimated carbon stocks

Fig. 6.1 (a) Conceptual diagram of baseline and project scenario carbon stocks for afforestation and reforestation projects, where project implementation leads to enhancement of carbon stocks. (b) Conceptual diagram of baseline and project scenario carbon stocks for avoided deforestation projects, where project implementation ensures that existing carbon stocks are maintained.

6.1.1 Fundamental Steps in Establishing a Baseline

Baselines can be characterized by making a projection of business-as-usual changes in land use and carbon stocks in the area where the project is proposed. The approach is to elaborate a scenario of possible future changes in land use and associated changes in carbon stocks under the "without-project" scenario. Usually this is done by considering past trends and the current situation and, based on this, making a projection. Two approaches to get the past trends for a project are: (i) compilation of historical data and (ii) participatory rural appraisal based on local community knowledge (Chapter 8). Key factors used in projecting the baseline have included planned land-use decisions of landowners and stakeholders, designation of land by national authorities and historical land-use change in the local area. Projections also have to take into account any events that may alter current practices or policies, such as changes in legislation related to land use, changes in market preferences or prices or changes in environmental awareness. Evidently, there are several risks no matter how thoroughly the task is handled. The four fundamental steps in establishing a baseline are as follows.

Step 1: Define the project area and the boundary and stratify the area into homogenous strata (see Section 10.3 for strata definition).
Step 2: Establish past trends in land-use systems.
Step 3: Estimate carbon stocks (see Chapter 10–13 for methods) in all the land-use strata for the base year and for at least one more point in time prior to the base year.
Step 4: Project the future land-use scenario and carbon stocks

6.1.2 Project-Specific and Generic Baselines

Several approaches and methods are available for determining the baseline. Two types of baselines commonly considered are: (i) project-specific and (ii) generic or regional. *The project-specific baseline* The procedure to establish a project-specific baseline, also defined as the bottom-up method (Moura-Costa et al. 2000), compares different land uses or management practices to identify those that best represent the baseline scenario. The carbon stock associated with the baseline scenario becomes the reference carbon stock level, which is compared to that after the project activity to calculate the gains in carbon stock due to the project. In following this procedure, all types of feasible land uses and management practices identified for the project area, and the associated carbon stocks, should be assessed. Because land use, management practices and changes in both are often spatially and temporally variable, a detailed project-specific study is likely to predict emissions more accurately than a broader regional or sector-based assessment. Project-specific baseline development involves assessment of future land use as well as changes in carbon stocks in the land-use systems in the without-project scenario. Approaches to and methods of developing project-specific baselines are presented in Chapter 7. The project-specific approach can be easily monitored and verified; however, it is limited by high cost and inconsistent methods adopted across projects in a region. The Executive Board of Clean Development Mechanism (CDM) under the Kyoto Protocol has approved several methodologies for developing project-specific baselines for afforestation and reforestation projects.

Project-specific baseline could prove problematic for four reasons. First, it is inherently difficult to predict the future. Secondly, under Clean Development Mechanism, project managers have strong incentives to overstate the decline in carbon stocks in the baseline scenario because such overestimates increase the predicted gains in carbon stocks and thereby the relevant revenues for the projects. Thirdly, baseline setting requires some assumptions about regional and national land-use related programmes and policies (Chomitz 1998). Finally, project-specific baselines have high transaction costs (Watson et al. 2000).

Generic or regional baselines Regional baselines offer an alternative to some of the limitations of project-specific baselines and can be considered as standardized baselines for larger reference areas, which can avoid high or low project-specific baselines. Regional baselines could be developed for larger homogeneous zones such as agroecological zones, forest types or other types of spatial characterization based on rainfall, soil, altitude, topography and vegetation cover, using the methods presented in Chapter 7. This baseline could be adopted by all project developers in the region. Such baselines enhance the transparency of methods, facilitate review of the methods and reduce the uncertainty introduced because of methods and subjective judgments. The benefits of a generic or regional baseline are low cost and suitability for projects of different sizes (from small scale to regional), while their drawback is lack of site-specific accuracy, the reason why this procedure finds only limited use in land-based mitigation projects.

Adoption of the project-specific baseline approach will require estimation of baseline carbon stocks and projection of changes in the stocks for each project. A generic or regional baseline, on the other hand, will require estimation of baseline stocks and their rates of change on per hectare basis for each land-use system: these values could be applied to any project in the region that constitutes a single agroecological zone, which may comprise several districts, states and provinces.

6.1.3 Fixed and Adjustable Baselines

A baseline can be determined by either assuming a fixed and steady state or using an adjustable or moving approach. The question to ask is whether the baseline established at the start of the project should remain constant throughout the project or be adjusted periodically, as described by the re-estimated baseline in Fig. 6.1a.

(i) A fixed-state baseline approach can be used where the carbon stock (line B(S)) is unlikely to change over the years (Fig. 6.1a). Highly degraded lands or agricultural lands with no change in use or management practice are unlikely to experience changes in carbon stocks. Further, any small changes in, say, soil organic carbon cannot be reliably measured. Thus, only a single measurement at the beginning of the project is adequate. Many CDM methodologies recommend the fixed baseline approach (http://cdm.unfccc.int).

(ii) If carbon stocks are expected to change under the baseline scenario, it is more appropriate to adjust the baseline periodically and estimate carbon stocks at different periods (Fig. 6.1a). Use control plots in a proxy area outside the boundary of the project but subjected to the same conditions as those that obtain within the project plots for monitoring changes in carbon stock. The central argument for revising the baseline over the length of the project is that such revisions may ensure more realistic estimates of carbon stock changes; the central argument against is that a continual revision could involve additional costs and make projection of future carbon stocks a complex process.

6.1.4 Baseline Scenario for Carbon Mitigation and Land-Based Development Projects

The extent of importance of a baseline, methods for estimation, carbon pools to be monitored and frequency of monitoring is different for carbon mitigation and other land-based development projects (Table 6.1). Baseline estimates are

Table 6.1 Baseline-related issues for carbon mitigation and other land-based projects

Issue	Carbon mitigation project	Land-based development projects
Importance	- Very critical to estimate the additional or incremental carbon benefit due to project activities	- Not critical for roundwood production projects involving planting trees - Soil carbon stock at the start of a project necessary to assess improvement in soil fertility
Methods	- Baseline carbon stocks could be fixed or adjustable - Need to adopt standard methods depending on the mechanism or agencies, e.g. - CDM: approved methodologies (http://cdm.unfccc.int) - GEF: Pearson et al. (2005a)	- No standard or prescribed methods - Standard textbook methods to measure growth or stocks of biomass or soil carbon - Involves measuring carbon stock at two periods
Pools	- All carbon pools that will be affected by project activities	- Roundwood production: only above-ground biomass - Land reclamation: only soil organic carbon
Frequency	- Periodically throughout the project period	- Only at the pre-implementation stage

required for all projects, but critical for carbon mitigation projects involving stringent methods, monitoring of all carbon pools and more frequent monitoring. Even a roundwood production project requires the estimation of a baseline, but the project also requires estimates of pre-project biomass stock or soil organic matter stock. Baseline estimation and monitoring for a carbon mitigation project involves higher transaction costs because of more frequent monitoring of several carbon pools.

6.2 Additionality (UNFCCC) or Incrementality (GEF)

Additionality and incrementality are concepts associated with estimating changes in carbon stock due to the project activities in relation to the estimated baseline. A project is expected to reduce CO_2 emissions or increase carbon stocks.

Additionality is the additional reduction in CO_2 emissions or gain in carbon stock – quantified relative to baseline levels – that would not have occurred in the absence of the project. Although a project activity is generally assumed to differ from its baseline scenario, the proposed project activity or a management practice it employs sometimes may have been implemented anyway. In such a case, the project activity and its baseline scenario are effectively identical, leading to no additionality. Additionality is defined by the UNFCCC for land-based CDM projects as follows: "The proposed afforestation or reforestation project activity under the CDM is additional if the actual net greenhouse gas removals by sinks is increased above the sum of changes in the carbon stocks in the carbon pools within the project boundary that would have occurred in the absence of the registered CDM project activity". It can be illustrated using Fig. 6.1 where additionality of carbon gain is the difference between P(R) and B(R) for afforestation (Fig. 6.1a) or (P–B) for avoided deforestation projects (Fig. 6.1b).

The important principle guiding funding by the Global Environment Facility (GEF) is that the project would not otherwise be undertaken without GEF funds, demonstrating the incrementality – a term used within GEF. GEF has provided guidelines for estimating incremental gains in carbon stock due to project implementation.

Estimation of additional or incremental changes in carbon stocks due to carbon mitigation projects requires carbon stock values for the baseline as well as project scenarios and calculation of the difference or change in carbon stocks. Estimation of additionality is also required for forest conservation, roundwood production and land reclamation projects. These non-climate-mitigation projects have less stringent requirements (Table 6.2): it is enough to estimate carbon stocks at the beginning of the project and compare them to the stocks in a future year.

Non-carbon, land-based projects also require estimation of total roundwood production or incremental soil organic matter content, without involving stringent methods.

Table 6.2 Additional or incremental carbon stock for carbon mitigation and other land-based projects

Issue	Carbon mitigation projects	Other land-based projects
Importance	- Estimation of additionality over a baseline is a requirement	- Estimation of above-ground biomass stock or soil carbon density at a given period is often adequate
Methods	- Approved and prescribed methods should be adopted, e.g. - CDM: unfccc.org/cdm - GEF: Pearson et al. (2005a)	- No standard or prescribed method - Standard textbook method to estimate merchantable timber or soil organic matter at a given period
	- Methods require estimates of carbon stocks for baseline scenario and project mitigation scenario	- Carbon stock at a given period, such as at end of rotation, is compared with pre-project stock
Pools	- All carbon pools are likely to be affected by project activities	- Mostly one carbon pool adequate - Roundwood production projects: above-ground biomass - Land reclamation projects: soil organic carbon
Frequency	- Determined by the mechanism or programme - CDM: once in 5 years - GEF: periodically	- At harvest time for roundwood production - At the end of the project

6.3 Permanence

The concept of permanence is associated with risk in general and with re-emission of carbon conserved or sequestered in particular (Pearson et al. 2005b). The idea of permanence is that carbon conserved or sequestered should be over a long term and in a sense permanent. This risk of loss of permanence can be due to natural events such as wildfires, which can be very hard to anticipate, or due to human actions. For example, reduced access to land, food, fuel and timber resources without any alternative resources may result in carbon loss due to reversal of project activities. Loss of permanence could involve, for example, deforesting the forest area conserved or harvesting trees raised under an afforestation programme and using them as fuelwood, leading to emission of CO_2. The final benefit to climate is greatly dependent on the actual carbon sequestered or CO_2 emissions avoided because of project activities over a given period. Therefore, non-permanence is an important parameter while estimating carbon gains. Non-permanence requires careful accounting of loss or extraction of carbon stocks. In principle, the approach that estimates changes in carbon stock between two points in time takes care of any loss of carbon stocks.

Non-permanence may not have direct implications for carbon inventory if the"Stock-Difference" approach is adopted (see Chapter 9 for information on the

Table 6.3 Permanence issue for carbon mitigation and other land-based projects

Issue	Carbon mitigation projects	Land-based projects
Importance	- Critical for mitigation projects, since carbon sequestered or conserved could be emitted back to the atmosphere	- Roundwood production projects: not critical, though sustainable harvesting is desirable - Land reclamation projects: loss of soil organic carbon has implications for soil fertility
Methods	- Carbon stock change method addresses the non-permanence issue - Has implications for accounting but not for measurement methods	- No implications for methods
Pools	- All carbon pools likely to be lost or reemi-tted to the atmosphere are relevant	- Soil organic carbon for land reclamation projects
Frequency	- Same as for estimating additionality	- Not relevant, though permanent soil organic matter or fertility improvement is critical

"Stock-Difference" approach). Non-permanence of carbon gains is an issue for carbon mitigation projects but not for other land-based projects such as roundwood production or agroforestry (Table 6.3), where any extraction or loss is accounted for in the estimates of roundwood production or soil carbon stocks.

6.4 Leakage

Leakage is defined as the unanticipated decrease or increase in carbon emissions or removal outside a project's accounting boundary (see Chapter 8 for more information on project boundary) as a result of project activities. It can be referred to as the offsite effect (Aukland et al. 2003; Sathaye and Andrasko 2006). For example, reforestation of an area used for grazing can lead animal owners to take the livestock to new land outside the project boundary for grazing. The type of leakage varies with the type of projects, but most land-based projects are subject to leakage. Occurrence of leakage requires assessments outside the project boundary (Ravindranath and Sathaye 2002).

There are different types of leakages (Table 6.4), which can arise from both market effects and shifting of activity and can have both positive and negative impacts. From a carbon inventory point of view, both positive and negative leakages are important although the negative impacts in terms of carbon mitigation or climate benefits are of greater concern.

Primary leakage occurs when carbon benefits of a project are entirely or partially negated by similar processes in another area. This means that the negative activity has moved elsewhere as a direct effect of the implementation of the project. Primary leakage can include processes such as extraction of fuelwood,

Table 6.4 Different types of leakages and their features

Type of leakage	Description
Positive spillover	Implementation of a project could lead to replication of activities such as forest conservation, sustainable extraction, reforestation and grassland reclamation practices outside the project boundary, possibly driven by socio-economic benefits, ultimately enhancing carbon storage outside the project boundary.
Shifting land conversion	Implementation of a project activity may stop land conversion within the project boundary but shift it to locations outside the project boundary. This will lead to carbon emissions outside the project boundary, which will be a negative leakage, negating the overall carbon benefit of the project.
Shifting source of biomass extraction or livestock grazing	Implementation of a project may lead to shifting of extraction of biomass (fuelwood or timber) and grazing to outside the project boundary. Such a shift in woody biomass extraction or grazing may lead to carbon emission outside the project boundary.

Table 6.5 Implications of leakage for carbon mitigation and other land-based projects

Issue	Carbon mitigation projects	Land-based projects
Project boundary	- Project boundary definition involves inclusion of areas likely to be affected by project activities, even if outside the project intervention area, for leakage estimation - Boundary needs to be defined for carbon inventory	- Project boundary usually restricted to area subjected to direct project activities - Area directly subjected to project activities is the project boundary for most projects, except forest conservation
Importance	- Critical for estimation of net additional carbon benefits and for all mitigation projects, most critical for avoided deforestation	- Roundwood production and land development projects: leakage not an issue - Forest conservation: a critical issue from the perspective of biodiversity, flow of forest products, loss of tree crown
Methods	- Approved methods to be adopted, e.g. - CDM: http://cdm.unfccc.int - GEF: Pearson et al. (2005a) - Methods require monitoring of carbon stocks in land area outside the project area	- No specific methods - Relevant only to forest conservation projects - Monitoring of non-project area impacted may be necessary
Pools	- All carbon pools, particularly soil organic carbon and above-ground biomass for projects involving land-use change	- Forest conservation: above-ground biomass and soil carbon if land-use change is involved
Frequency	- Periodicity similar to that for monitoring carbon pools	- Monitoring at the end of the project may be adequate

timber and grazing in another area and it can also occur because of shifting of land conversion or biomass extraction outside the project boundary, leading to CO_2 emissions.

Secondary leakage occurs when a project's output creates incentives to increase carbon emission elsewhere. Unlike primary leakage, secondary leakage activities are not directly linked to or carried out by someone directly involved in the project. If project implementation leads to oversupply of timber, causing a reduction in prices in the region and increased timber consumption because timber is cheap, it can be regarded as secondary leakage.

Leakage estimation is an integral component of carbon mitigation projects and to some extent of forest conservation (non-climate-related) projects. Biodiversity conservation or watershed protection could be the focus of forest conservation or protected area management projects. Conservation in project area should not lead to deforestation or land conversion outside the project boundary. Leakage estimation may not be relevant to roundwood production, agroforestry and land reclamation projects. The implications of leakage to carbon mitigation and other land-based projects are presented in Table 6.5. The project boundary is critical for assessment of leakage in both carbon mitigation and other land-based projects.

6.5 Project Boundary

The project boundary is the geographically delineated area dedicated to the project activity. Projects can vary in size from hundreds of hectares to hundreds of thousands of hectares either as a contiguous unit or distributed as multiple parcels under a single project management. The spatial boundaries of the land need to be clearly defined and properly documented for measurements and monitoring. Defining project boundary is necessary to estimate the leakage of carbon benefits due to implementation of the project. A project area can have a primary boundary and a secondary boundary:

- *A primary project boundary* is the geographic boundary restricted to areas, locations and land-use systems directly subjected to project intervention or activities such as protection, management and planting.
- *A secondary project boundary* may have to be delineated and marked to include locations and land-use systems outside the project boundary impacted or experiencing leakage due to shifting of land conversion or biomass extraction or livestock grazing. This area can also be subject to the actual project activity.

Both boundaries should be identified and described for carbon mitigation projects. The choice of an accounting boundary influences the carbon stock gains that are assigned to a project. The project boundary is critical for carbon mitigation projects, incorporating the need to include area outside the boundary to assess the impacts on carbon stocks. This issue is not relevant to other land-based projects, except forest conservation projects (Table 6.5).

6.6 Scale of the Project

The size of a project determines the methods to be used for carbon inventory: carbon stock changes in small-scale projects could be monitored using field measurements whereas large-scale projects may require adoption of remote sensing and modelling techniques. Small-scale projects are likely to be more homogeneous with respect to soil, topography, species dominance and silvicultural practices than large-scale projects, which are likely to be heterogeneous, requiring multistage stratification. The heterogeneity or homogeneity of a project also determines the methods to be adopted for boundary determination, stratification, sampling and selection of carbon pools.

Small-scale projects, which are likely to be more homogeneous, can be adequately served by simple methodologies, which reduce the cost of estimating carbon stocks under baseline and project scenario. Small-scale afforestation and reforestation projects are included in this category and recommended default factors can be used for assessing the existing carbon stock taking into account issues such as soil, lifetime of project and climatic conditions (Sanz et al. 2005).

The concept of bundling refers to the idea of combining a number of small and similar projects into a single monitoring system to reduce the transaction cost associated with monitoring. Bundling is permitted under the Clean Development Mechanism (CDM) projects, which means that projects may be bundled as long as the total size is below the limit for a single CDM project (Lee 2004).

Implications of the scale of project for carbon mitigation and other projects Scale of the project may not impact the carbon pools to be selected or periodicity of monitoring for carbon mitigation or other land-based projects.

6.7 Conclusions

Carbon inventory for land-based carbon mitigation and other conservation and development projects is influenced by several methodological issues with implications for the inventory: (i) large heterogeneity of soil, topography and vegetation types with implications for rates of carbon stock changes; (ii) need for a baseline scenario to determine additional carbon stocks; (iii) non-permanence of carbon stock gained; (iv) leakage of carbon stock gained within the primary project boundary; (v) need for estimating the net additional or incremental carbon stock gain over the baseline stock gain or loss; (vi) scale of the project with implications for heterogeneity of land-use systems; and (vii) project areas directly impacted by the project activities and areas indirectly impacted by the project activities.

These factors have implications for the methods to be adopted for the carbon inventory (field measurement or remote sensing or modelling), stratification and sampling (single or multistage sampling), selection of carbon pools (single or multiple) and frequency of monitoring. The contentious issues in land-based carbon mitigation projects addressed in this chapter are baseline development, additionality of carbon benefits, non-permanence of carbon stock gains, leakage of carbon benefits, project boundary, and scale of the projects. It was shown that these issues have implications for carbon inventory methods for carbon mitigation projects. However, issues such as baseline and permanence are not of much relevance to forest, grassland and agroforestry development projects. These other land-based conservation projects aim at roundwood production or soil reclamation or biodiversity conservation. A summary of implications of various methodological issues for carbon mitigation and other land-based conservation and development projects is given in Table 6.6.

Table 6.6 Implications of different methodological issues for carbon mitigation and land-based conservation and development projects

Issue	Carbon mitigation projects	Land-based projects
Baseline	- Very critical for estimating net carbon benefits - Approved methodologies to be used - Requires periodic monitoring of relevant carbon pools if carbon stocks are dynamic	- Not very critical for assessing roundwood production or soil fertility - Estimates of carbon stocks at the beginning of the projects adequate
Additionality	- Estimation of additional carbon stock gains over the baseline carbon stock changes is necessary - Approved methodologies to be used - Periodic monitoring of carbon pools in project and baseline scenario necessary - Multiple carbon pools are relevant	- Carbon stock estimates at a given period such as at the end of rotation or project - Standard textbook methods adequate - Need to monitor only one or two carbon pools
Leakage	- Estimation of leakage of carbon benefits outside the area subjected to direct project activities necessary for estimating net carbon benefits - Approved methodologies to be used	- Leakage relevant to forest conservation projects, if protection in one area leads to forest conversion in another area
Permanence	- Estimation of reversal or loss of carbon benefits required - Carbon Stock-Difference method estimates any loss due to reversal of carbon	- Not an issue for most land-based projects
Project boundary	- Includes areas directly subjected to project activities as well as areas not directly subjected to project activities but where carbon stocks will be impacted	- Not an issue for most land-based projects, except forest conservation projects
Scale	- Has implications for carbon inventory methods and cost of monitoring	- Has implications for carbon inventory methods

Chapter 7
Carbon Inventory Under Baseline and Project Scenarios

The main goal of carbon inventory for carbon mitigation projects is to estimate the incremental or additional biomass and soil carbon stocks gained because of project activities. Estimation of incremental or additional carbon benefit or stock gain requires monitoring of carbon stocks of a given area and over a given period under the "without-project" or "baseline scenario" as well as changes in stocks for the project area over the same period as a result of implementation of project activities under the "project scenario". Carbon inventory methods are required in a project cycle during project proposal development, post-project implementation and monitoring phases. This chapter describes the broad approaches, methods and steps for estimating carbon stock changes at two levels:

- Under baseline and project scenarios
- At project development and monitoring phases of the project cycle

These approaches and methods are also applicable to non-carbon land-based projects, such as roundwood production. Detailed methodologies for measurement, calculation, projection and monitoring of different carbon pools are described in the later chapters.

7.1 Broad Approaches to Estimating Carbon Stocks

Three broad approaches to estimation of carbon stocks and changes under baseline and project scenarios are those based on:

(i) Default value
(ii) Cross-sectional field studies
(iii) Modelling

7.1.1 Approach Based on Default Value

Default values used in estimating carbon stocks are values obtained from databases, published literature or previous studies of similar land-use systems in similar environments. Default values of carbon stock and growth rate are available in the literature (IPCC 2003, 2006, www.fao.org and www.efdb.int). The approach based on default values does not involve any field studies and measurements of biomass and soil carbon pools but involves the following steps:

- Defining the physical and biological conditions of the project location, namely rainfall, soil type and topography
- Defining the land-use category, namely forest land, degraded forest land, degraded grassland, cropland, shrub land and plantations of different species
- Identifying the vegetation type, namely forest type, plantation type, dominant tree or grass or shrub species, age of the stand and crop grown
- Identifying management practices such as grazing, thinning, fertilizer application and irrigation
- Defining the land category for which default data are required, e.g. land-use category, rainfall, soil type, gradient, forest type or plantation species, age of the stand, density of trees and grazing practice
- Searching published literature and databases as well as local or regional studies or measurements conducted earlier
- Selecting relevant carbon pools
- Selecting an appropriate default value for the carbon pool for the land category defined for the project location, based on expert judgement

Baseline scenario Default data are required for biomass and soil carbon pools for degraded forest, cropland and grassland areas proposed for, say, afforestation, which are subjected to different levels of grazing, biomass extraction and degradation. Very few sets of default values are available for such land-use conditions for the baseline scenario. Avoided deforestation projects may require default biomass and soil carbon data for the forest stand; these values are, however, available.

Project scenario Default values for stocks and growth rates of above-ground biomass, below-ground biomass, soil carbon, litter and deadwood are available for forests, plantations, grasslands and agroforestry systems. The stocks and growth rates of different carbon pools vary not only among different land-use categories (e.g. forests, grassland and cropland), vegetation types (e.g. grassland, eucalyptus plantations and savannah), but also significantly within a given vegetation type (e.g. evergreen forests, savannah, eucalyptus plantation) under different rainfall, soil and temperature regimes. Carbon stocks or growth rates may vary even for a given tree plantation (e.g. eucalyptus or pine) or agroforestry system, depending on management practices, such as irrigation or fertilizer application. Therefore, default values should be evaluated considering the physical (soil and rainfall) and vegetation conditions (e.g. dominant species and variety) and management practices (fertilizer or irrigation application). Management practices vary often, and the default values may not be available for a given species-mix or soil condition and, even when available, may not be suitable for a given location, such as a village forest or a patch of grazing land. Expert judgement is required in using a default value from a global or regional database. The uncertainty could be very high if the default value is not available for location-specific conditions. Default values are required, particularly during the project development stage, for calculating carbon stock in both baseline and project scenarios.

7.1.2 Approach Based on Cross-Sectional Field Studies

Default values on carbon stocks in different carbon pools for different land-use categories and subcategories, which are required for baseline and project scenarios, may not be available often. Further, default values of carbon stocks for lands subjected to different periods of degradation are unlikely to be available in literature, and those for rates of change in carbon stocks (above-ground biomass and soil carbon) in degraded, barren, or fallow lands are even less likely to be available. Therefore, most often project developers may have to generate carbon stock data on their own for the land-use categories and land area selected for project activities, particularly under the baseline scenario.

Baseline scenario The approach to estimating carbon stocks based on cross-sectional field studies makes use of field and laboratory measurements. Adopting this approach would require access to land-use categories such as cropland, grassland and degraded lands that are subjected to changes similar to those the proposed project lands are subjected to under the baseline scenario and with comparable rainfall, soil, topography, vegetation cover and management practices.

Project scenario Cross-sectional studies involve identification of locations with desired vegetation, soil, topography, management systems and age class for the project activities. At most project locations, forests or plantations of different age and subjected to selected management practices are likely to be available, such as eucalyptus plantations or naturally regenerated forest plots of different age classes promoted under ongoing or past afforestation and reforestation programmes in the region. Plots with similar vegetation characteristics relevant to the project activities may be located even outside the project boundary with similar soil, rainfall and other conditions. If such plots are available, carbon stocks and changes can be estimated based on the field and laboratory methods described in Chapters 10–13. The values for carbon stocks or growth rates thus obtained could be used for the proposed project during the project development phase to calculate or project future carbon stock gain, roundwood production or soil quality improvements.

7.1.3 Approach Based on Modelling

Models can be used at various phases of the project cycle to make projections of future carbon stock changes. Models are particularly useful during the project development phase, when estimates of future carbon stocks resulting from project activities are required at the project design or proposal stages. Models such as PRO-COMAP, the project-based comprehensive mitigation assessment process (Sathaye and Meyers 1995), CENTURY (CENTURY 1992) and CO_2-FIX (Schelhass et al. 2004) can be used to make projections of future carbon stocks for a given carbon pool based on the base year carbon stocks and estimated rates of change for a given land use or vegetation type (refer to Chapter 15 for details of these and other models). Some models such as CENTURY and CO_2-FIX are process-based models, requiring additional data on various ecological, physiological, soil and water parameters. The reliability of the projections depends on the quality of input data as well as the capacity of the model to simulate carbon dynamics.

Baseline scenario The PRO-COMAP model calculates future baseline carbon stock changes of different pools based on the base year carbon stock and the rate of change defined for each of the carbon pools, particularly above-ground biomass and soil carbon.

Project scenario PRO-COMAP, CO_2-FIX and CENTURY models could be used for projecting carbon stock changes under the project scenario involving implementation of project activities aimed at enhancing the carbon stocks or avoiding CO_2 emissions.

Models require estimates of stocks and rates of change in the stocks of different carbon pools for making projections. These values can be obtained from default value data sources or values measured as part of field studies.

7.2 Carbon Inventory Under Baseline Scenario

7.2.1 Selection of Fixed or Adjustable Baseline

Carbon inventory under the baseline scenario involves estimation and projection of changes in stocks of different carbon pools in the project area at project development and monitoring phase. It is possible to visualize two situations with respect to baseline carbon stock changes with implications for carbon inventory: the stocks may change, declining normally under the baseline scenario, or may remain stable over the period under consideration. The concept of baseline and baseline scenario is explained in Chapter 6.

Fixed carbon stocks under baseline scenario The carbon stock in the baseline scenario may have stabilized over the years and is unlikely to change significantly during the project period. For example, the land use or management practices on degraded forests, grasslands and croplands may not have changed over the years, leading to stabilization of carbon stocks. Thus, the carbon stock needs to be measured only for the project base year, the assumption being that the stocks would remain stable over a given period in the future. Adoption of this approach reduces the cost of measuring carbon stock changes over the years. This approach has been adopted for some of the approved Clean Development Mechanism (CDM) methodologies (http://cdm.unfccc.int) under the Kyoto Protocol. Soil carbon is likely to undergo minimal or no measurable change under the baseline scenario for a majority of land-use systems over short periods of 5–10 years. Further, litter and deadwood biomass stocks are likely to be either absent or present in insignificant quantities that do not merit estimation. Thus, above-ground biomass pool is likely to be the only pool that may change in degraded lands. However, even above-ground biomass is unlikely to undergo any change in land-use categories such as degraded forest, grassland and cropland. Carbon stock is likely to undergo minimal or no change in the baseline scenario under the following situations:

- Forest land remaining forest land with no major change in extraction or management practice
- Degraded forest land remaining degraded forest land with a few or no regenerating trees or shrubs and with no disturbance to topsoil
- Grassland remaining grassland with no major change in livestock grazing density and disturbance to topsoil
- Cropland remaining cropland with no major change in cropping system, e.g. annual cropland remaining annual cropland or perennial cropland remaining perennial cropland
- Barren or degraded forestland, grassland and marginal cropland unlikely to show any measurable change in carbon stocks unless topsoil is disturbed

Dynamic or adjustable carbon stocks under baseline scenario Carbon stocks could change over the years due to change in land use or management practice or even due to changes in the intensity of use and management practices. Changes in carbon stocks could be drastic because of practices such as ploughing that disturb topsoil or gradual because of continued grazing. The likely situations leading to changes in carbon stocks under baseline scenario are as follows:

- Forest land converted to degraded forest land
- Forest land or degraded forest land converted to cropland
- Grassland converted to cropland
- Perennial cropland converted to annual cropland
- Cropland converted to agroforestry
- Fallow land ploughed and cropped

The extent of change in stocks of carbon pools is likely to depend on the land-use category and changes in land use and management practices. For example, above-ground biomass is likely to change significantly because of conversion of forest land to other categories of land use. Soil organic carbon and below-ground biomass stocks are likely to undergo significant changes when forest or degraded forest or grassland is converted to cropland or managed grassland involving intensive land preparation and management. Litter and deadwood will be significantly impacted when forest land is converted to managed grassland or cropland.

However, according to some of the approved CDM methodologies, baseline carbon stocks have to be measured and reported even when the changes are insignificant or not measurable or within an error range (http://cdm.unfccc.int).

Selection of fixed or adjustable baseline This is the first step in estimating carbon stocks and projecting changes in them under the baseline scenario. The selection of the type of baseline has implications for carbon inventory estimation methods as described in Sections 7.2.2 and 7.2.3. The selection would be based on expert judgement regarding the likely changes in carbon stocks in the future under baseline scenario conditions. If land use or management practices are expected to change, impacting carbon stocks, adopt an adjustable baseline. If an adjustable baseline is selected, the carbon stocks will have to be measured or estimated periodically. If the land-use system or management practices have stabilized and if the land is too degraded with no likely changes in carbon stocks in the future, adopt a fixed baseline, requiring estimation only once at the beginning of the project.

7.2.2 Baseline Carbon Stock Estimation at Different Phases of Project Cycle

Estimation of carbon stocks and changes under the baseline scenario in the absence of project activities is necessary to evaluate carbon gains resulting from implementation of the project activities. Since the baseline is a projection of carbon stocks in the project area in the absence of project activities, stocks of different carbon pools may increase, decrease or remain stable under a given land-use system and management practice if no project is implemented.

Estimation of carbon stocks and changes in the stocks for the baseline scenario in such carbon mitigation projects as afforestation and reforestation under CDM is determined by the approaches and methods prescribed by its Executive Board (http://cdm.unfccc.int). Land-based conservation and development projects not aimed directly at carbon mitigation also estimate carbon stocks and changes in the absence of project activities. In the following sections, broad approaches and steps for estimating carbon stocks and changes are presented for land-use projects aimed at carbon mitigation as well as for other kinds of projects. The approaches to and steps in estimating and monitoring carbon stock changes under the baseline scenario vary with:

- Project development phase
- Post-project implementation or monitoring phase

In the project development phase, approaches to estimating baseline carbon stocks could be based largely on literature or default values and, in a few cases, on cross-sectional studies. However, during the project monitoring phase, the estimation approaches are largely based on field and laboratory studies.

7.2.3 Baseline Carbon Stock Estimation and Projection During Project Development Phase

The project development or pre-implementation phase requires estimation and projection of changes in carbon stocks over the selected project area and period in the "without-project" scenario. This is also referred to as *ex ante* calculation of projected changes in carbon stocks. These projected stocks and changes need to be reported at the project development phase to estimate and project the impact of the proposed project activities on carbon stock gains.

7.2.3.1 Approach 1, Based on Default Value

The following steps could be adopted for *ex ante* calculation of carbon stock changes in the baseline scenario using the default value-based approach:

Step 1: Define the project boundary covering all the parcels of land to be brought under the project activities (refer to Chapter 8).

Step 2: Stratify the project area into homogeneous land classes based on tenure, soil, topography and baseline vegetation status under the baseline scenario conditions before the implementation of the project (refer to Chapter 8).

Step 3: Stratify the project area by overlaying the homogeneous land classes obtained in step 2 with the proposed project activities (e.g. planting different species, different grazing practices and different forest management practices).

Step 4: Define and demarcate the strata (see Section 10.3 for strata definition) dedicated to different project activities based on step 3 for the base year (t_0) incorporating the current land-use status (step 2) and proposed project activities (step 3) and estimate the area under each stratum such as:

- o Stratum 1 comprising degraded grassland proposed for short-rotation afforestation but subjected to fuelwood extraction
- o Stratum 2 comprising degraded grassland proposed for short-rotation afforestation but subjected only to regulated grazing
- o Stratum 3 comprising cropland proposed for long-rotation afforestation

Step 5: Select the carbon pools relevant to each of the stratum defined in step 4.

Step 6: Estimate the carbon stocks for all the selected strata under the baseline conditions for the base year t_0, using default values available from other studies, reports and programmes in the region or from a published database.

Step 7: Select one of the following two approaches to estimation and projection of carbon stock change under the baseline scenario:

- o Fixed carbon stock
- o Adjustable carbon stock

Step 8a: If the fixed-carbon stock approach is used, estimate the stocks of different carbon pools only once for the base year t_0, assuming that the stocks will not change over the project period

[or]

Step 8b: If the adjustable-carbon stock approach is used, estimate the carbon stocks at different selected periods for different pools using default values for the change in carbon stocks.

Step 9: Based on current and historical land-use data and any ongoing or proposed programmes for the project area, project future land-use systems for different periods, for example, 5, 10, 15 and 20 years for each stratum.

Step 10: Use future land-use pattern for a selected year, for example, t_5, t_{10}, t_{15}, t_n and use the default values for carbon stocks

Step 11: Estimate carbon stock for the future period (5–10 or 20 years or t_5, t_{10}, t_{20}, respectively) for all the land strata defined in step 4 using default data for the soil carbon and above-ground biomass carbon pools.

Step 12: Calculate the difference between the carbon stocks considering all project land-use systems and area for the year t_n (projected period) and year t_0 (base year, the project starting date) using the following formula:

Change in carbon stock in the baseline or without-project

$$\Delta C = Ct_n - Ct_0$$

scenario (ΔC)

where ΔC = change in carbon stock in tC/ha, C_{tn} = carbon stock in year t_n (tC/ha), C_{t0} = carbon stock in base year t_0 (tC/ha). The value of ΔC could be positive or negative, but is most often negative, indicating reduction in carbon stocks.

7.2.3.2 Approach 2, Based on Field Measurement of Carbon Stocks through Cross-Sectional Studies

The procedure involves the same steps as those described for Approach 1, except that the carbon stock values for the current and projected land-use systems are obtained by field measurements through cross-sectional studies. Carbon stocks should be estimated for the base year t_0 and a projected future year t_n.

(i) *Carbon stocks for the base year t_0* could be estimated using the following steps during the project development phase:

Steps 1–5: are identical to those in the default value method described earlier to identify and demarcate different land strata.

Step 6: Using methods and procedures given in Chapters 10–13 (for different carbon pools), estimate the total carbon stock for year t_0 for each land stratum in the project area based on measurements using "plot method", described in Chapter 10.

(ii) *Carbon stocks for the future year t_n* could be estimated using the following steps during the project development phase. This approach is necessary only if changes in land use or management projected under the baseline scenario include those such as degraded forest or grassland converted to cropland:

Step 1: Derive the future land-use systems and areas for each of the stratum under the baseline scenario based on historical data, participatory rural appraisal (PRA, Chapter 8) and any ongoing or proposed programme for the time period selected (t_5, t_{10}, t_{15}, t_n).

Step 2: Select relevant carbon pools for the future land-use systems, which may be similar to or different from the pools for the current land-use system strata.

Step 3: Obtain future carbon stock data for each projected land-use system by identifying land areas subjected to conditions leading to the new land-use system for the period t_n:

- o Locate areas that have experienced the projected land-use changes (e.g. forest land converted to grassland or cropland) or changes in management practices (e.g. grazing) within the project boundary or nearby areas outside the project boundary for the period selected
- o Estimate carbon stocks in areas subjected to land-use or management practice change in the past using methods and procedures given in Chapters 10–13
- o Estimate and calculate total carbon stocks considering the projected land-use systems and area

Step 4: Estimate change in carbon stocks in the baseline scenario using the following procedure:

- o Estimate total carbon stock for base year (year "t_0")
- o Estimate total carbon stock for a future project year such as t_5, t_{10} or t_{20} using the steps described above
- o Estimate the change in carbon stock between the future project year and base year using equation provided in Section 7.2.3.1

7.2.4 Baseline Carbon Stock Monitoring During Monitoring Phase

7.2.4.1 Situations Requiring Periodic Monitoring

Monitoring of stocks of different carbon pools is required if these are expected to change under the baseline scenario or in the "without-project" situation. It is necessary to identify the factors driving changes in land use or management practices under the baseline scenario that contribute to carbon stock changes. Simulation of such factors is necessary for monitoring during the period identified for the project scenario. Potential factors that may lead to changes under baseline scenario are:

- Need for additional land for food production or livestock grazing
- Increase in livestock population and grazing intensity
- Increase in demand for fuelwood
- Changes in cropping practices

Changes in carbon stocks could be driven by a single factor or a combination of factors. Often, it may not be easy to delineate the factors driving land-use change, a management practice, or a land-use pattern or to link these factors to impacts on carbon stocks.

The rates of change in carbon stocks may also depend on the status of vegetation and soil in the base year. If the land-use system has high carbon density, any small change in management practice, such as degree of harvesting or land preparation or livestock grazing density, may lead to large changes in carbon stocks. Potential situations with differing impacts for carbon stock changes are as follows:

- *Forest land with high biomass and soil carbon density* is likely to undergo significant changes in carbon pools following any change in management or land-use practices, such as rate of extraction of biomass or disturbance to topsoil.
- *Degraded forest land with low carbon density* (both soil and biomass) is likely to undergo marginal change in carbon pools, particularly if topsoil remains undisturbed.
- *Degraded grassland with no tree cover* is likely to undergo no change or only marginal changes, particularly if topsoil remains undisturbed; however, disturbance to topsoil will lead to significant loss of soil carbon.
- *Cropland with no tree cover* is likely to undergo marginal carbon stock changes, especially if agricultural practices remain unchanged.
- *Cropland with agroforestry and significant tree cover* is likely to undergo changes in carbon stocks if the tree cover or soil is disturbed, but not if agricultural practices remain unchanged.

Thus, it is important to consider the following factors before deciding on monitoring in the post-implementation phase:

- Whether the area is subjected to land-use change
- Whether the area is subjected to any change in management practice
- Whether carbon stock in the land – above-ground tree biomass and soil carbon – is high or low

Periodic monitoring of carbon stock changes may not be critical if the lands are not subjected to changes in use or management practice and if the carbon stocks are very low.

7.2.4.2 Control Plot Approach for Carbon Stock Monitoring under Baseline Scenario

Monitoring of carbon stock changes under the baseline scenario during post-implementation project period requires simulation of the "without-project" conditions. It is necessary to simulate the baseline conditions to monitor the periodic changes in carbon stocks of different pools. It may be difficult to simulate such "without-project" conditions as grazing or extraction within the project boundary since the project developers are interested in bringing maximum potential area under the project activities. Therefore, the "control plot" approach could be adopted to monitor changes in carbon stocks, which involves delineating plots of land of required sample size and allowing "without-project" scenario conditions such as grazing or removal of fuelwood and leaves to persist. Two approaches could be adopted for laying out control plots:

(i) *Control plots within the project boundary* Control plots could be established within the project boundary for long-term monitoring of carbon stock changes. Refer to Chapter 10 for size and number of control plots to be established for long-term monitoring. Normally, plots of 0.25 ha with four replicates are adequate. Whereas large control plots exceeding a hectare may not be feasible, plots of 0.25 ha dispersed in different locations within project boundary are an alternative. Four replicates of 0.25 ha each, totalling 1 ha, account for only 0.1% of a 1,000 ha project area. Normally, most projects will have multiple parcels of land. The control plots could then be distributed among different land parcels of the total project area. The plots should be accessible for grazing or fuelwood extraction, similar to "without-project" or "pre-project" conditions.

(ii) *Control plots outside the project boundary* Control plots could be located outside the project boundary when it is not feasible to do so within the project boundary, for example when the project area is very small or forms a single contiguous plot. Control plots could be located adjacent to the project boundary or even in nearby villages or grazing lands or watersheds subjected to similar "without-project" scenario conditions. Locating control plots outside the project boundary may not be difficult in most situations, since similar land-use categories as proposed in the project would be available. Further, factors driving carbon stock changes such as grazing, fuelwood extraction and land-use change commonly occur over a large landscape. Refer to Chapter 10 for details on size and number of control plots to be located outside the project boundary.

The broad steps in monitoring carbon stock changes involve estimating the area under different land-use systems, establishing project boundaries, locating control plots and measuring carbon stocks in different carbon pools of the control plots.

Step 1: Adopt steps 1–4 described in Section 7.2.3.1 and identify and demarcate the strata selected for the project along with estimates of area under each stratum.

Step 2: Based on historical records of land-use change, the participatory rural appraisal method and current land-use pattern, project future land-use pattern for the project area under baseline scenario.

Step 3: Establish project boundary for the project (refer to Chapter 8).

Step 4: Identify the drivers that contribute to changes in land-use and management practices impacting carbon stocks.

Step 5a: Establish permanent control plots for monitoring carbon stock changes for each land-use system in the project area that will not be subjected to project activities but are likely to be subjected to conditions or drivers similar to those under the "without-project" scenario

[or]

Step 5b: Establish such permanent control plots outside the project boundary.

Step 6: Identify the carbon pools likely to be impacted by the drivers leading to land-use change or changes in management practices.

Step 7: Measure and estimate carbon stock changes in the control plots for the carbon pools identified using the methods given in Chapters 10–13 and the "permanent plot" technique.

Step 8: Convert carbon stock values from sample plots to per hectare values

Step 9: Using per hectare data on changes in carbon stocks and area involved, estimate the total baseline carbon stock change for the selected period.

7.2.5 *Baseline Estimation Through Use of Models*

Carbon stock estimation and projection could be made using models as described in Section 7.1.3. Adoption of models such as PRO-COMAP or CO_2-FIX also requires data on carbon stocks and rates of changes. Input data for the models could be obtained from default data sources or from field measurements. The models help in projecting carbon stocks for a given period. The steps involved include the following:

Step 1: Select the land strata.
Step 2: Select a model suitable for the project activities.
Step 3: Identify the input parameters required for making projections, e.g. baseline biomass and soil carbon stock and rate of change under the baseline conditions, and area of the stratum.
Step 4: Generate the input parameters by adopting the default value approach as well as cross-sectional field studies. Refer to the relevant earlier sections in this chapter for the approach and steps to be adopted for generating input parameters through default and cross-sectional approaches.
Step 5: Input the parameters into the model and generate future carbon stocks or incremental gain or loss for a given project activity and area.

7.3 Carbon Inventory Under Project Scenario

The project scenario involves planning, implementation and monitoring of a set of project activities aimed at achieving the project goals of gaining carbon stocks (mitigation projects), roundwood production or soil quality improvement. All these projects require carbon inventory, either directly (for carbon mitigation projects) or indirectly (for other land-based conservation and development projects). This section presents the broad approaches to carbon inventory for both carbon mitigation and other land-based projects for the project scenario for:

- Project development phase
- Project monitoring phase (post-implementation phase)

7.3.1 *Project Development Phase*

Most land-based projects involve activities that conserve existing carbon stocks in a given land-use system or increase carbon stocks through implementation of project activities. Estimation and projection of carbon stocks and changes during the project development phase are made using one or a combination of the following approaches:

(i) Default values
(ii) Cross-sectional field studies
(iii) Modelling

The approaches in the project scenario refer to carbon stock estimates resulting from implementation of project activities.

(i) *Approach based on default values* The approach based on default values is most relevant at the project development phase. In the project scenario, project activities to be implemented to achieve project goals, such as carbon mitigation or forest conservation and roundwood production, are selected and described. These activities could include planting trees and improving practices related to harvest or grazing or soil conservation, which lead to conservation of carbon sinks or enhancement of carbon stocks. Default values for carbon stocks or rates of change available in literature for selected activities could be used. The steps to be adopted for *ex ante* calculation of changes in carbon stock in the project scenario are identical to those in the baseline scenario (Section 7.2.3.1):

- Select the project activities to be implemented, e.g. establishing plantations of eucalyptus or pine or other tree species or reclaiming grassland or conserving forests
- Identify local soil, rainfall and topography conditions
- Select the carbon pools and refer to literature or databases or other local project reports for default values for the selected land-use category and carbon pools

Use the steps given in Section 7.2.3.1 and project future carbon stocks to be used in the project proposal.

(ii) *Approach based on cross-sectional studies* The approach based on cross-sectional studies can be used during the project development phase to estimate future carbon gains for a given future project year. The approach is likely to provide more reliable estimates of carbon stocks than those provided by the default value-based approsach. The approach based on cross-sectional studies is described in detail in Section 7.2.3.2 and the key steps to be adopted are as follows:

Step 1: Select all the project activities considered for implementation in the project along with the species and management practices (e.g. natural regeneration through regulated grazing, planting eucalyptus at a density of 2,000 plants/ha with no fertilizer application or irrigation).

Step 2: Conduct a reconnaissance survey of the region or participatory rural appraisal (see Chapter 8 for the method) or refer to relevant land-based programmes implemented in the region. Collect all relevant information on the location, area brought under the activity, year of initiation and silvicultural practices adopted.

Step 3: Select locations with characteristics similar to the proposed project activity such as species planted, density, fertilizer application or irrigation, and mark them on a map showing the year in which the activity was implemented.

Step 4: Adopt the stratification, sampling, measurement and calculation procedures described in Chapters 10–13 for the selected carbon pools to estimate carbon stocks and growth rates.

Step 5: Estimate and project the gains in carbon stock to be achieved by implementing the proposed project activities using the area under each project activity and the rate of change in carbon pools obtained from cross-sectional studies.

(iii) *Approach based on modelling* Models are particularly relevant to make projections during the project development phase for the project activities. Adoption of models such as PRO-COMAP, CO_2-FIX and CENTURY requires generation of input data for making the projections using default data or those obtained from cross-sectional studies. Select the model and adopt the steps given in Section 7.2.5 to make projections of carbon stock changes.

7.3.2 Monitoring Carbon Stock Changes in the Project Scenario

Monitoring of carbon stock changes is required in the post-project implementation stage for all projects, whether carbon mitigation or other land-based conservation and development projects. The monitoring phase starts only after project activities are implemented and carbon stock gains begin to accrue in quantities significant enough to be measured and estimated. Unlike the project development phase, the monitoring phase is largely based on field and laboratory studies. Impacts of the project are measured during this phase in relation to project goals. One of the key initial steps is to identify the carbon pools impacted by the project activities for monitoring. Monitoring of different carbon pools starts at different periods depending on the carbon pool and can continue for long periods, even decades, depending on the features of the project activity. Refer to Chapter 4 for details on selection of carbon pools and frequency of monitoring. The methods to be adopted during the monitoring phase are:

(i) Field and laboratory studies
(ii) Remote sensing
(iii) Modelling

The suitability of these methods to estimate carbon stocks and changes in the stocks for different broad project types is given in Table 7.1. All project types require field and laboratory methods involving measurements. Measurement methods generate input data as input to the models for projections and for validating the results obtained from remote sensing. Remote sensing methods are particularly useful in monitoring changes in land use or land cover over the years while models can be useful in projecting carbon stocks based on input data on carbon stocks and growth rates, which is described in Chapter 15.

(i) *Field measurement and laboratory studies* Land-based projects in forest, cropland and grassland categories are characterized by large diversity of vegetation, soil, rainfall, topography, altitude and management practices. These differences make large impacts on carbon stocks and growth rates. Further, these physical and biological factors have differing implications for stocks and rates of changes on different carbon pools. For example, grasslands have more of soil organic carbon than biomass carbon, and valley lands have higher soil carbon than sloping lands do. Thus, the key carbon pools and methods for carbon inventory will vary with soil, rainfall, temperature and topography apart from vegetation types

Table 7.1 Suitability of field studies, remote sensing and modelling for estimating different carbon stocks under different types of projects

Project type	Field studies	Remote sensing	Modelling
Avoided deforestation	- Biomass and soil carbon stocks and changes - Area changes	- Distinct land-use changes - Canopy cover change	- Not suitable for area estimation - Requires data on carbon stocks
Afforestation, reforestation and bio-energy plantations	- Area planted annually from project authorities - Permanent and control plots for monitoring biomass and soil carbon stocks	- Monitoring distinct land use or land cover changes	- Not suitable for area estimates or monitoring - Equations available for projection of biomass
Forest management	- Area under management from project authorities - Measurements of carbon stocks	- Not suitable for area estimation - Not suitable for carbon stock change estimation	- Not suitable for area estimation and biomass and soil carbon estimation
Agroforestry	- Area from project authorities - Permanent plots for monitoring biomass and soil carbon stocks	- Area and canopy change estimation feasible - Biomass and soil carbon estimation not feasible	-Area monitoring, biomass and soil carbon projection not feasible
Grassland reclamation	- Area from project authorities - Biomass and soil carbon from permanent plots	- Area monitoring not feasible - Biomass and soil carbon estimation not feasible	- Area estimation not feasible - Biomass and soil carbon estimation limited
Shelter belts	- Area from project authorities - Biomass and soil carbon from permanent plots	- Area estimation feasible - Estimation of biomass and soil carbon not feasible	- Area estimation not feasible - Biomass and soil carbon estimation limited

and management practices. Given the diversity and variation, the best approach to monitoring carbon stock pools is to conduct field studies at the project location to measure biomass and soil carbon pools, since this will yield actual location-specific data. Field and laboratory studies have been routinely adopted for monitoring forest, roundwood, grass and crop production projects apart from soil organic carbon monitoring in all land-use categories. Field and laboratory studies for monitoring different carbon pools are described in detail later (Chapters 10–13) and involve the following broad steps:

Step 1: Define the project boundary encompassing all land where project activities have been implemented and demarcate it on a map.

Step 2: Monitor and record land-use changes and implementation of project activities in the project area annually.

Step 3: Stratify the project area based on the following:

- Project activity, such as forest area brought under protection, raising a plantation, grassland management, different agroforestry systems and shelterbelts
- Any critical differences in soil, topography or management

Step 4: Identify key carbon pools for different project activities and strata.

Step 5: Adopt the "permanent plot" method for periodic measurement of carbon pools in the selected strata in the project area.

Step 6: Select an appropriate method for each carbon pool (Chapters 10–13).

Step 7: Estimate the change in carbon stocks of different carbon pools according to the frequency of monitoring selected for each pool and aggregate all the pools for the project area for the period under consideration.

Step 8: Monitor leakage of carbon stocks outside the project boundary through control plots (Chapter 6).

Step 9: Monitor changes in carbon stock under the baseline scenario (refer to Section 7.2).

Step 10: Calculate the net gain or loss in carbon stocks due to the project activity and for the total project area over the period selected.

Net additional/incremental carbon benefit for the selected period = (carbon stock change under project scenario – carbon stock change under the baseline scenario – leakage)

(ii) *Modelling for gains in carbon stocks* Very often, project managers want to project future carbon benefits based on the initial measurements to get an idea of the likely long-term gains in carbon stocks. Modelling techniques can be used to project such gains monitored over a given initial period, 5 years, for example, into the future, may be 10 or 30 years after the project implementation. Models such as PRO-COMAP, which use average carbon stock and rate of change for each pool, can be used to make projections with limited input data:

Step 1: Select the project activities and strata.

Step 2: Select the model suitable for the project activity, for example, PRO-COMAP modules for short rotation and long rotation or CO_2-FIX for projecting carbon stock changes in a forest regeneration project.

Step 3: Identify the input parameters required for making projections.

Step 4: Generate the input parameters based on field measurements and laboratory studies during the initial years and use default data where relevant.

Step 5: Input the parameters into the model and generate future carbon stocks or incremental stock gains for a given project activity and area, based on the monitoring carried out for the initial period.

(iii) *Remote sensing techniques* Remote sensing techniques could be used for monitoring distinct changes in area under different vegetation types, such as area under forest with closed canopy, degraded forest with open canopy or grasslands within the project boundary. Monitoring changes in area using remote sensing techniques is particularly useful in large projects or in assessing the impacts of project activities on land-use systems outside the project boundary. Remote sensing techniques could also be used for monitoring forest tree canopy cover, leaf area index and other parameters to estimate the aboveground biomass. Refer to Chapter 14 for detailed discussion on the steps and methods as well as the benefits of adopting remote sensing techniques for monitoring area and changes in carbon stocks.

7.4 Summary of Methods

Estimates of gains and losses in carbon stocks are required at different phases of a project cycle, particularly project development (pre-project implementation) and monitoring (post-project implementation) phases. The broad approaches to and steps required for monitoring, estimation and projection of changes in carbon stocks for carbon mitigation or other land-based conservation and development projects are described in this chapter. Multiple approaches, such as field measurement studies and modelling, are required for any typical project. Use of default data may be necessary at all phases and particularly during the project development phase. Detailed methods for measurements, monitoring, calculation and projection of carbon stock changes are presented later in this handbook.

Chapter 8
Techniques for Estimation and Monitoring of Project Areas and Boundary

Estimating the area and delineating the boundary of a land-use category for greenhouse gas inventory and of area dedicated for a proposed project is the first basic step in preparing a carbon inventory. Methods ranging from field measurements, such as a physical land survey, to more complex methods involving the use of a satellite available for estimating land area and marking its boundary are presented in this chapter. The need for stratification of land area to increase the accuracy will be highlighted (methods for stratification are described in Chapter 10). These methods could be used in preparing greenhouse gas inventories of different land-use categories for use in carbon mitigation or roundwood production projects. Estimating the project area and the boundary is very critical for carbon inventory since this determines the following:

(i) The total project area directly impacted by the project activity.
(ii) The area from which leakage of carbon stocks occurs and can be estimated.
(iii) The area from which areas sample plots for carbon inventory can be taken.
(iv) Whether the project boundary is fixed or changing (the boundary can change over time due to additional area added to the project activity or new areas subjected to leakage).
(v) Whether the project activity comprises a single contiguous unit or dispersed multiple units.

These features will have implications for sampling and for the cost of preparing the carbon inventory. Any change in project area may require additional sampling units, and multiple parcels dispersed over the project area may need more sampling plots. Describing and listing the features of land uses and land cover of the area can serve as the basis for estimating land areas used in a project. It is good practice to define a land-use system at a disaggregated level to enable its stratification into homogeneous subsystems. The land-use types could, for example, be described as forest land, cropland, grazing land, barren land or grassland. These types can be further disaggregated, such as forest land being split into natural primary forest, degraded forest, eucalyptus plantation or regenerated forest. Each one of these different land-use systems or subsystems of a project can be found as a single unit or in multiple units dispersed over different areas, such as villages or landscapes. These land-use systems and subsystems should be presented spatially on a map.

It is possible to visualize two situations: one in which information and maps are available for the land-use category or project area describing the land-use and cover features, vegetation status and latitude and longitude. The other in which not much information is available, and even a map may be lacking. For some locations, satellite maps may be available. Therefore, the extent of effort involved in defining the area and boundary depends on the status of preliminary information availability and the access to it.

Broad methods Two broad methods are available for land area measurement and boundary marking:

1. Ground methods
   (a) Physical measurements
   (b) GPS approach
   (c) Participatory rural appraisal
2. Remote sensing methods
   - Aerial photography
   - Passive and active satellite imagery

Practical issues *in choosing a method* Selection of a method depends on several factors such as size of the project, resources available and technical capacity, and these factors should be evaluated in making a choice.

- *Type of the project* Different types of land-use projects will demand different kinds of methods to get the required information.
- *Size of the area* Size is critical because large projects extending to tens of thousands of hectares may need remote sensing techniques for monitoring changes in the area, while small, village-scale community projects may be adequately served by measurements in the field.
- *Level of accuracy* Coarse-resolution satellite data will give generalized information over each unit at a pixel level, whereas fine-resolution satellite data offer greater details leading to greater accuracy in interpretation.
- *Technical capacity* Trained field staff is needed for ground measurements whereas computer hardware and software expertise is needed for making the best use of satellite imagery.
- *Cost* Resources available determine whether aerial photography or physical measurements are used for a given project, small or large.
- *Physical access* Remote areas in dense forests or high up in the mountains are hard to reach, which makes field measurements difficult, and remote sensing methods may be the only choice.

8.1 Approach to Selecting a Method

Project area estimation and boundary marking is the first step in project development as well as implementation. Estimation of the project area as well as its composition, that is whether it comprises a single unit or multiple units, is required for estimating the total carbon stock gains. The approaches to selection involves the following steps:

Step 1: Identify the land-use category or land-use systems and the location selected for the project, along with the size of the project.

Step 2: Obtain all the maps available for the project location such as a topographical map, soil map, land-use map and maps derived from remote sensing data.

Step 3: Collect information related to current and historical land-use patterns, land tenure, human settlements, livestock grazing locations, source of fuelwood and timber and locations and areas of implementation of different programmes such as afforestation, soil conservation and grassland reclamation.

Step 4: Collect detailed information on proposed project activities such as area planned for afforestation, soil conservation and forest protection and phasing of these activities, particularly the area to be brought under project activities annually and locations of different project activities.

Step 5: Evaluate all the currently available maps and information as well as the details of the proposed project activities along with the resources available for monitoring to take a final decision on the method to be adopted for measurement and monitoring of project area and activities.

Examples of methods for different land-use categories and project activities are presented in Table 8.1. Normally ground methods are used for small-scale and dispersed location projects in landscapes consisting of a mosaic of multiple land-use

Table 8.1 Methods for estimating and monitoring of project area and boundary for different land-use categories and project activities

Land-use category	Scenario	Status of land and vegetation	Scale of project: Small scale	Scale of project: Large scale
Forest land	Baseline scenario	- Natural forest converted to degraded forest or other land uses	PRA	Remote sensing
	Project scenario	- Preserving natural forest	Physical measurements	Remote sensing
Grassland	Baseline scenario	- Grassland remaining grassland and subjected to degradation	PRA	Physical measurements
	Project scenario	- Grassland reclamation - Afforestation	Physical measurements GPS	Physical measurements Aerial photography
Cropland	Baseline scenario	- Cropland remaining cropland	PRA	Physical measurements
	Project scenario	- Agroforestry	Physical measurements GPS	Remote sensing
		- Afforestation	Physical measurements GPS	Aerial photography

systems. Aerial photography and remote sensing techniques are used for large-scale projects with adequate financial and technical resources. Remote sensing techniques are still evolving, particularly for application in small-scale projects. Participatory rural appraisal techniques could be effectively used to supplement other methods particularly for small-scale projects in and around human settlements.

8.2 Ground Methods

Ground methods refer to methods involving deployment of field personnel on the ground in the area that is investigated. Ground methods or field surveys can produce accurate and detailed data, which is their strength. On the other hand, the methods are often time consuming, expensive and demand skilled personnel, especially in remote areas and harsh terrains. Thus, these methods are suitable for smaller areas and require intensive preparation for detailed measurement of area and other features (Rosillo-Calle et al. 2006). Three methods are presented and they can be used separately or in combination to increase the accuracy of the estimation of land area.

8.2.1 Physical Measurement

This ground method involves physical measurements of the units or parcels of land on the ground using measuring tapes. The material and preparation needed are as follows.

Materials required include a compass, measuring tape or chain (30–100 m), hammer, survey pegs, permanent-marker paint, clinometer (ocular device showing changes in percent and in degrees), shovel and notebook

Preparation for fieldwork Good knowledge and information on existing data from the area, a person with local knowledge in the team, permission from landowners and land users obtained by explaining the work's purpose and maps available for the area.

If the land is leveled, the area can be measured by using a tape or chain and compass. If the site has a significant slope (greater than 10 degrees), a clinometer should be used to record the slope in degrees. This is done by taking a clinometer reading from one position to the next, ensuring that the next location is at the same height as the operator's eyes. If the slope is greater than 10 degrees, a correction factor has to be used for estimating the distance, which can be done with a correction table given in Table 8.2. Two people make the work easier (Greenhouse 2002).

The choice of a distance-measuring device should be evaluated based on where it will be used. Dense or thorny environments will destroy a regular measuring tape and thick vegetation will entangle it, hence a chain is required. In open grassland areas, a measuring tape made of fibreglass is preferable, since a fabric tape may expand since it is slightly elastic and should therefore be avoided. Metal tapes are available but usually in short lengths.

Table 8.2 Distance slope correction table

Slope percent	Slope distance per 30 m
5	30
10	30.1
15	30.3
20	30.8
25	31.1
30	31.6
35	31.9
40	32.5
45	33.1
50	33.8
55	34.4

8.2.1.1 Steps in Physical Measurements

Step 1: Hold or tie the measuring device at a fixed point along the border of the land area. Tag the starting point and all the points on the boundary whenever the direction changes for periodic monitoring. Tagging can be done with metal clips that are added to trees or other permanent features.

Step 2: Take a bearing to a point on the boundary in the direction in which you will be walking.

Step 3: Walk towards the point where the bearing was taken letting the measuring device unfold as you walk.

Step 4: At each change of direction, record the covered distance using the measuring device.

Step 5: Check the previous bearing with a back bearing and record it.

Step 6: Take a new bearing to the next point.

Step 7: If the slope is greater than 10 degrees, measure this in degrees with a clinometer at every change of direction.

If available, digitalize the result into a geographical information system (GIS) since it can be easily stored, changes incorporated and linked to other data for the project. Map, either manually or digitally, the spatial distribution of different parcels of land, locations of the parcels, project activities along with the boundary and extent of each parcel of land.

8.2.1.2 Merits and Demerits of Physical Measurement

Merits

- Provides ground knowledge of the project area to the project managers
- Easy to adopt and local staff could be trained to implement the method
- Suitable for projects covering small geographical areas
- Suitable for project proposal development phase

Demerits

- High cost, particularly for large areas
- Difficult if there are multiple land units far apart
- Not suitable for large sized projects

8.2.2 GPS Approach

The Global Positioning System (GPS) can be used to estimate land areas, either by walking along the boundary for small projects or by traveling in a vehicle for areas larger than that for ground physical measurements described earlier (Section 8.2.1). GPS is being extensively used in all research in land-based projects such as ecological studies and agricultural development.

GPS is defined as the constellation of satellites that circle the Earth in a very precise orbit and transmit signal information to the Earth. GPS receivers take this information and use triangulation to calculate the user's exact location. Essentially, the GPS receiver compares the time a signal was transmitted by a satellite with the time it was received. The time difference tells the GPS receiver how far it is from the satellite (www.navcen.uscg.gov). Any reading of a position is accompanied by information on which satellite is used in the calculation and the accuracy of the reading.

Single unit GPS is what is referred to as a handheld GPS, and uses several satellites to pinpoint a location. A more precise GPS system is to use a reference station on the ground, which can be a permanent unit of a portable unit. This is referred to as differential GPS. The method described here is based on a single unit, handheld GPS, which gives an accuracy of ±10m. This accuracy is adequate for most land area estimations.

GPS data of the boundary of a land area can be stored in a geographic information system and facilitate future verifications (Greenhouse 2002). The positions can also easily be applied manually to any map that shows the latitudes and longitudes.

Materials required include GPS, batteries, a tagging device, notebook and a local map showing latitudes and longitude for verification.

Preparation for fieldwork Inform landowners and land users of the objective of the project and involve key informants who can provide information on the current and past land-use patterns.

8.2.2.1 Steps for GPS

Step 1: Set up the GPS according to the manufacturer's instructions. Determine the geodetical datum, coordinate system and units required and enter them into the GPS. Use the same parameters as the end product, usually the base map used for the project. If there is no information on any of these parameters, a basic set-up is World Geodetic System (WGS84) for geodetical datum, Universal Transverse Mercator (UTM) or Plane for coordinate system and metres/feet for unit. Most GPS have these parameters prelisted and can therefore be easily selected.

Use a local map that gives latitudes and longitudes.

Step 2: At the starting point of GPS readings, let the GPS obtain an average reading over 5 min to obtain the most accurate reading. Usually a minimum of four satellites are required to get a reasonably accurate location.

Step 3: Check your position on the map. If there are differences, check the datum settings and try again. If the accuracy is not obtained, move to a new site and try for a new reading. Dense crown cover and hills can lower the accuracy.

Step 4: For periodic monitoring, tag the starting point and all points on the boundary whenever the boundary changes direction. Tagging can be done with metal clips that are added to trees or other permanent features.

Step 5: Make notes of error estimates given by the GPS (usually given on a relative scale or in metres).

If the GPS data are stored digitally it can usually be downloaded and used in GIS programs. If these possibilities are not available, the readings can be recorded manually. The land area can be calculated and then used in estimation and monitoring procedures. Accuracy of the positions given by a GPS is ±10 m, which for many planned projects is sufficiently good for land estimation. A handheld GPS unit can be bought starting from less than US$100 and can go up to a few thousand dollars.

8.2.2.2 Merits and Demerits of GPS Technique

Merits

- Supplies data even while on the move, that is, the data are not restricted to a few fixed locations
- Adequate accuracy (±10 m)
- Compatible with GIS
- Suitable for projects covering a small geographical area
- Suitable particularly for project proposal development phase

Demerits

- In dense forests or mountainous regions, signals from the satellites can be obstructed, lowering the accuracy of data
- Difficult if there are multiple land units that are far apart

8.2.3 Participatory Rural Appraisal

Participatory rural appraisal (PRA) as a method has developed mainly through developmental work among non-governmental organizations (NGOs) and agencies, spurred by the need to find methods and techniques for effective interaction with communities (Mikkelsen 1995). The PRA is characterized by flexibility and learning on the site through face-to-face interaction. PRA tools relevant to land-based projects include semi-structured interviews with individuals or groups, participatory mapping with transect walks and different types of time line, trend and seasonal diagramming. One of the benefits of PRA is the additional information generated that usually has direct relevance to understanding land-use patterns and changes.

The idea is to facilitate community members in developing spatial representations of their area by creating maps that can be used with the existing information (Asia Forest Network 2002). The maps that the participants produce could reflect the location of villages, forests, agricultural land, water resources as well as management issues and ownership.

It is important to prepare adequately before conducting a PRA to ensure adequate participation and information generation. The following guidance would be useful:

- Establish rapport with the stakeholders
- Involve stakeholders and community members depending on land resources and who are knowledgeable about the existing and past land-use systems and factors driving change and current dependence on land
- Enlist at least two people to conduct the exercise

8.2.3.1 Steps in Conducting PRA

Step 1: Organize a group meeting at a time convenient to the participants at the project site.

Step 2: Explain the purpose of the PRA exercise and the project.

Step 3: Hold group discussions using a set of questions but be flexible to depart from the set if the responses demand such deviation.

Step 4: Let the facilitator present the issues or raise questions and make notes and his or her assistant record the responses.

Step 5: Guide the participants in drawing a map of the project area depicting different land-use systems on the ground, on paper or on transparencies laid over an existing map. The facilitator can have a checklist of land-use types likely to be included and should guide the participants to depict them on the map.

Step 6: Walk with the participants to cover the project area they have described. Tagging can be done with metal clips added to trees or other permanent features or GPS readings can be taken to record the route.

Step 7: Transfer mapped information onto a paper or an existing map or digitize the maps for GIS analysis.

Step 8: Calculate the areas under different land-use systems, historical changes and likely projected changes.

8.2.3.2 Merits and Demerits of Participatory Rural Appraisal

Merits

- No instrumentation or expensive material required.
- Valuable additional information, particularly about the historical changes in land-use pattern, dependence on the land area and likely land-use patterns in future.

- Involvement of local communities.
- Suitable for projects covering a small geographical area.
- Useful tool to use throughout a project cycle.

Demerits

- Data can be subjective
- Difficult if project area comprises multiple land units far apart
- Not very useful if the participants have limited dependence on or knowledge about the project location

8.3 Remote Sensing Methods

Generally remote sensing can be defined as the science of observation from a distance. Data in remote sensing are acquired by sensors mounted on an airplane or optical and radar sensors on board a satellite or from cameras equipped with optical or infrared films installed in aircraft. These data are usually classified to provide estimates of the land cover and its corresponding area (IPCC 2003). Classification can be done either by visual analysis of the imagery or by digital and computer-based methods. Each method presents advantages and disadvantages. More information on remote sensing for carbon inventory is given in Chapter 14.

To synthesize information relevant to a particular area or region, the technique of overlaying maps with different sets of information such as soil, topography and land-use pattern is adopted through Geographic Information Systems (GIS), which use computer hardware and software (Barret and Curtis 1995). GIS is also used for image processing and interpretation. Digital classification of data obtained from remote sensing can also be adopted for GIS, which allows the data to be manipulated in different ways, such as merging different spectral data into indices. Digitization also allows immediate computation of areas associated with the different land-use categories.

If data obtained from remote sensing are interpreted digitally in a GIS, a number of corrections are needed. Atmospheric correction will account for aerosols, water vapour and smog. Geometric correction will account for angle of the sensor and refer the image to some sort of coordination system. Many products available today in several databases (see Chapter 14 for details) are pre-corrected by the producer. Remote sensing has developed rapidly over the past decade, along with associated technical computer development, making hardware, software and also the satellite data readily available at low cost in most countries.

Examples of well-established software are ArcView, IDRISI, PCI, ERDAS, ERMAPPER and ENVI. ArcView is a commercial product, which is vector based, which means that its images are composed of lines, polygons and points. The IDRISI software is a product from Clark University, which is raster based, which means that the images are composed of pixels, which is similar digital remote sending data. Both packages can run on a normal PC, although storage is problematic because digital images are large; the problem can be solved by using an external hard disk.

Supervised and unsupervised classification There are several classification techniques to be performed in the estimation of land area. Unsupervised classification refers to the use of an algorithm and ground truth data to classify an image. The ground truth data indicates the positions of several pixels with a particular land-use type and the image values give a range for this land use. The program then classifies the whole image using this information, assigning all the pixels with the corresponding range of values to the same category as that assigned on the basis of ground-truth information. Usually the range of values for a particular land use overlaps those for other land-use type, yielding ambiguous results. In cases where several spectral bands are available for the satellite data (see Chapter 14 for more information), unsupervised classification can be made on a combination of bands. This enables several classifications to be compared, the assumption being that areas assigned to the same category in several classifications are more likely to be the dominant land-use types. However, because human supervision is lacking, the procedure is prone to more errors.

Supervised classification refers to a two-step classification analysis with greater human involvement in the classification process. Some unsupervised classification is used, followed by incorporating more ground-based data into the classification. The locations may be visited again if required. Although such classification is more time-consuming and costly, the resulting maps are more accurate.

Remote sensing and ground-truthing Data from remote sensing need to be accompanied by "ground-truthed" parameters, irrespective of the resolution or source. This means that the data have to be validated against something empirical. Remote sensing analyses always contain interpretations, which need to be validated with other information to make them more accurate. Land-use maps can be used to aid the interpretation process.

Criteria for selecting remote sensing data and products Data from remote sensing should meet the following conditions (IPCC 2006):

- It should be possible to obtain adequate land-use type categorisation from the data.
- The data should be available at an appropriate spatial resolution to facilitate categorization by land use.
- The data should be available at an appropriate temporal resolution for estimating land-use conversions.
- Data from other sources should be available, against which the accuracy of interpretation from remote sensing data can be verified.

8.3.1 Aerial Photography

Aerial photographs are photographs taken from a distance, from an airplane for example. Cameras on airplanes are the simplest and oldest form of aerial sensors used to monitor features of land surface. The spectral resolution of cameras is usually very coarse, which makes aerial photography more useful when finer spatial data are more important than spectral information. Aerial photography is therefore useful for purposes requiring a finer scale, such as areas ranging from a few hundred to a few thousand hectares (Das and Ravindranath 2006).

Successive photographs along a flight strip include overlapping of up to 60% and create stereoscopic areas. This overlapping area can be used to form stereo pairs, and hence three-dimensional images. Structures such as forest tree species and their age and health or agricultural crops and agroforestry systems can be revealed from aerial photography with some training.

The materials required for using this technique include aerial photographs, a georeferenced map and plastic film.

8.3.1.1 Steps for Manual Aerial Photography

Step 1: Find several points that are visible both in the map and the aerial photograph.

Step 2: If the photograph does not have a scale, calculate it with the use of the photograph and map information. This is required if land area is to be estimated. Calculate the distance between two points in the map. Calculate the distance between the same two points in the aerial photograph.

$$\text{Scale} = \frac{\text{Ground(m)}}{\text{Photograph(m)}}$$

Step 3: Do visual interpretation of the land-use systems of interest on an overlaid plastic film.

Step 4: Calculate the land areas of selected land-use systems.

Aerial photography can be scanned and used digitally in a GIS.

8.3.1.2 Merits and Demerits of Aerial Photography

Merits

- High resolution
- Easy to interpret visually
- Good for land areas that are inaccessible or large
- May already be available for the required area, even in a time series
- Suitable for project development and monitoring phases

Demerits

- For some area, availability of aerial photographs is limited and the cost is high
- Absence of georeferencing makes digitalization difficult
- Includes tilts and errors
- If not available, it requires a relatively long time to obtain the photographs
- Not appropriate in rugged terrain
- If there are multiple land units far apart, more images are required
- Different stages of vegetation growth/degradation can be difficult to distinguish

8.3.2 *Passive Satellite Data*

Passive satellite data are based on sensors relying on reflectance of solar energy back to the sensor or detector. Different types of indices can be developed to facilitate the interpretation of images and estimation of different land areas. It is also possible to obtain time-series data for the desired land-use systems, since satellites cover the land-use systems of any project location continuously and regularly.

The smallest unit to be identified depends on the spatial resolution. At a spatial resolution of 30 m, units as small as 1 ha can be identified (IPCC 2006). More information on passive satellite data and steps for project area and boundary estimation is given in Chapter 14.

Material required includes images (purchased or downloaded) and computer software. Further, knowledge of digital image processing is required.

8.3.2.1 Merits and Demerits of Passive Satellite Data

Merits

- Available as time series covering long durations
- Applicable for land areas that are inaccessible or large
- Suitable for project monitoring phase

Demerits

- Fuzzy boundaries between land-use systems can be hard to distinguish
- Different stages of vegetation growth/degradation can be difficult to distinguish
- Clouds can disturb the image
- If there are multiple land units that are far apart, more images are required

8.3.3 *Active Satellite Data*

Compared to data from optics-based satellite described above, radar data are obtained from active sensors: the detector sends out a signal and measures the time and manner of the returning signal. Different surface qualities lead to differences in the returned signals and therefore the radar detector is able to distinguish surface textures. The main advantage of radar data is that the signals penetrate cloud cover, a common problem in the tropical regions.

A common type of radar data are obtained from the Synthetic Aperture Radar (SAR) system, which operates at microwave frequencies. By using different wavelengths and different polarizations, SAR systems may be able to distinguish between categories of land cover or between the biomass content of different vegetation (IPCC 2006).

More information on active satellite data and steps for project area and boundary estimation is given in Chapter 14.

Materials required include images (purchased or downloaded) and computer software. Application of the method requires knowledge of digital image processing.

8.3.3.1 Merits and Demerits of Active Satellite Data

Merits

- Not affected by clouds and can be recorded even during night hours
- Suitable for land areas that are inaccessible or large
- Particularly suitable for the monitoring phase

Demerits

- Fuzzy boundaries between land-use systems can be hard to distinguish
- Different stages of growth/degradation can be difficult to distinguish
- Technical requirements and costs are high
- If there are multiple land units which are far apart, more images are required

8.4 Estimating and Monitoring Land-Use Change

Monitoring of land-use change is required both for national greenhouse gas inventory as well as for climate change mitigation projects. Estimation and monitoring of land-use change requires comparison of land-use patterns between at least two points in time. The detail of land-use systems has to be set at a level in line with the objectives of the project. If a forest area is affected by encroachment and degradation, categorization between natural forest, degraded forest, regenerated forest, agricultural land and grazing may be necessary, while a project for agroforestry may need categorization between only two, namely agroforestry and agricultural land (IPCC 2006). When the categories are assessed at different points of time in terms of maps, a grid counting method can be adopted (Fig. 8.1). To adopt this method, the two maps

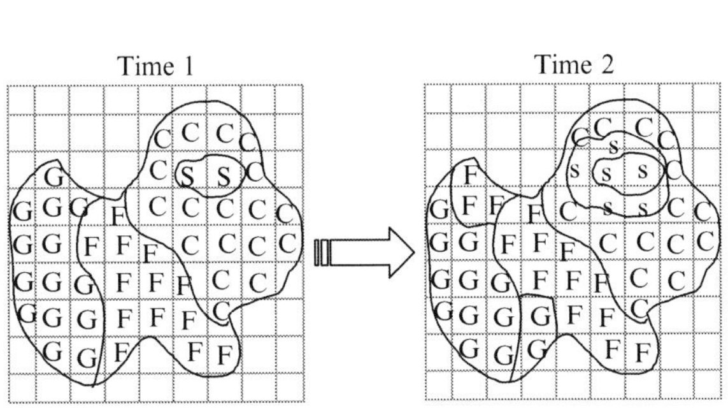

Fig. 8.1 Grid counting method for estimating land-use change between two points in time. (Modified from IPCC 2006.) F = forest land, G = grassland, C = cropland, S = settlements

Table 8.3 A simplified land-use conversion matrix. (Modified from IPCC 2006.)

Net land-use conversion matrix					
Time 1 / Time 2	Forest land	Grassland	Cropland	Settlements	Final sum
Forest land	11	3			14
Grassland	2	11			13
Cropland			15		15
Settlements			4	5	9
Initial sum	13	14	19	5	51
Numbers represent area units (ha)					

should have the same scale. The grid size should be fine enough to cover the smallest parcel of every distinct land-use type and preferable in a useful unit size, such as 100 × 100 m or 10 × 10 m.

The land-use system that dominates the grid will determine the land-use for that grid. Use the land-use conversion matrix (Table 8.3) for quantifying the change by counting the number of grids categorized as forest land at the two points in time. In the case shown in Fig. 8.1, the number of grids categorized as forest land decreased from 13 to 11 over the time period. The two grids (each representing one ha in this case) of forest land have been converted to grassland. On the other hand, three grasslands grids were converted to forest land during the same period.

The land-use conversion matrix in Table 8.3 shows the number of unchanged land-use systems in bold. Total areas as well as changes from one category to another are also shown in Table 8.3.

8.5 Conclusions

Estimation of area under different land-use categories and project activities is required for national greenhouse gas inventory as well as all land-based projects. Further, land-use change between two periods also needs to be monitored. Estimation of land area is required at project development phase, and periodic monitoring is necessary during the post-implementation phase. Two broad approaches presented in this chapter are ground measurement and remote sensing. The most common method adopted at project level is the physical ground measurement method. This is a low-cost and practical method for small-scale project areas. Remote sensing techniques are increasingly being used for large projects currently and may soon be available for small-scale projects. Application of remote sensing techniques is being simplified for use in most locations. The low-cost GPS method can also be used for estimating the project area.

Chapter 9
Generic Methods for Inventory of Carbon Pools

A carbon inventory involves estimation of changes in the stocks of the carbon pools over 1 year or between two points in time. Carbon inventory requires measurement and monitoring of all the selected carbon pools relevant to a land-based project or national greenhouse gas inventory for land-use categories. The features of different carbon pools described in Chapter 4 showed the need for different methods for inventory of different pools. Because multiple methods are available even for a single carbon pool, an appropriate method should be selected. The choice depends on several factors:

- Land-use category or land-use system, including vegetation cover
- Size of the land-use category and project area
- Accuracy of the estimate needed
- Resources available and cost-effectiveness of the method
- Project cycle; project development phase or project monitoring phase
- Technical capacity and institutional infrastructure available for inventory

In this chapter, the generic approach to estimating carbon through "Gain–Loss" or "Stock-Difference" method is presented (IPCC 2003, 2006) along with an overview of the various methods for estimating different carbon pools.

9.1 Approaches to Estimating Carbon Stock Changes

Carbon stock change is the sum of changes in stocks of all the carbon pools in a given area over a period of time, which could be averaged to annual stock changes. A generic equation for estimating the changes in carbon stock for a given land-use category or project is given below:

Annual carbon stock change for a land-use category is the sum of changes in all carbon pools:

$$\Delta C_{\mathrm{LUi}} = \Delta C_{\mathrm{AB}} + \Delta C_{\mathrm{BB}} + \Delta C_{\mathrm{DW}} + \Delta C_{\mathrm{LI}} + \Delta C_{\mathrm{SC}}$$

where ΔC_{LUi} is carbon stock change for a land-use category, AB = above-ground biomass, BB = below-ground biomass, DW = deadwood, LI = litter and SC = soil carbon.

The equation requires the stock change to be estimated for each of the pools. The changes in the carbon pool could be estimated using the two approaches based on IPCC guidelines (IPCC 2003, 2006):

1. Carbon "Gain–Loss"
2. Carbon "Stock-Change" or "Stock-Difference"

9.1.1 Carbon "Gain–Loss" Method

The carbon "Gain–Loss" method involves estimation of gains in carbon stock of the pools due to growth and transfer of carbon from one pool to another pool, e.g. transfer of carbon from the live-biomass carbon pool to the dead organic matter pool due to harvest or disturbance. The method also involves deducting losses in carbon stocks due to harvest, decay, burning and transfer from one pool to another as described in the following equation

Annual carbon stock change in a given pool as a function of gains and losses ("Gain–Loss" method) is given by:

$$\Delta C = \Delta C_{\mathrm{G}} - \Delta C_{\mathrm{L}}$$

where
ΔC is annual carbon stock change in the pool, ΔC_{G} is the annual gain of carbon, and ΔC_{L} is the annual loss of carbon.

The "Gain–Loss" method requires estimation of gain in the stock of each relevant carbon pool during the year or over a period under consideration in a given area. Similarly, losses in the stock of each pool need to be separately estimated and aggregated for a given area over a given period. The difference between carbon gain and loss will give an estimate of net carbon emission or removals.

9.1.2 Carbon "Stock-Difference" Method

The "Stock-Difference" method includes all processes that bring about changes in a given pool. The carbon stocks are estimated for each pool at two points in time, namely t_1 and t_2. The duration between the two points could be 1 year or several years, say 5, 7 or 10 years. As discussed in Chapter 4, the frequency of measurement of most of the carbon pools is once in several years – 5 years, for example, for soil carbon. Thus, the estimated stocks at t_2 need to be deducted from the estimated stock at t_1 and the difference divided by the number of years between the two periods (t_2 – t_1). The "Stock-Difference" must be estimated separately for each carbon pool.

Carbon stock change in a given pool as an annual average difference between estimates at two points in time (Stock-Difference method) is given by:

$$C = \frac{\left(C_{t_2} - C_{t_1}\right)}{\left(t_2 - t_1\right)}$$

where ΔC is the annual carbon stock change in the pool, C_{t1} the carbon stock in the pool at time t_1, and the carbon stock in the same pool at time t_2.

Changes in carbon stock using this method are estimated for a given land-use category or project area as follows:

- Estimate the stock of a pool at time t_1 and repeat the measurement to estimate the stock at time t_2.
- Estimate the change in the stock of selected carbon pool by deducting the stock at time t_1 from that at t_2.
- To obtain the annual change in stock, divide the difference in stocks by the duration $(t_2 - t_1)$ in years.
- If the estimates are made for sample plots, extrapolate to per hectare basis.
- To obtain the total for the project area, extrapolate the per hectare estimate to the total project or land-use category area.

The periodicity or frequency of measurement varies from pool to pool. Therefore, add the annual changes in each pool to obtain the total change in carbon stocks for the total project area over the selected period. It is important to ensure that the area under project activities between the two periods is identical; if it has changed, it is important to account for the change by calculating the changed area and multiplying it by the per hectare values.

9.1.3 Comparison of "Gain–Loss" and "Stock-Difference" Approaches

Of the two approaches, the "Stock-Difference" approach may be more suitable for estimating carbon stock differences for carbon mitigation as well as land conservation and development projects because of the following reasons:

- "Gain–Loss" approach requires estimation of rates of growth and losses of carbon pools, which can be obtained through "Stock-Difference" approach.
- It is difficult to estimate the losses due to extraction, fire, decay, burning and other causes in the project area.
- "Gain–Loss" approach requires apportioning the annual transfer of biomass into litter, deadwood and soil carbon pools, which requires significant additional effort.
- In the "Stock-Difference" approach, it is easier to account for changes in the stocks of all the relevant pools to obtain the per-hectare change although the frequency of measurement is different for different pools.

IPCC (2006) concludes that the "Gain–Loss" method is the default method, to be used when limited measured data are available. Further, the "Stock-Difference" method is suggested for greater accuracy. Accordingly, this handbook focuses on the "Stock-Difference" method.

9.2 Methodological Options for Estimating Carbon Pools

The project developer or manger and greenhouse gas inventory compiler will have to decide on the method to be adopted for carbon inventory of different pools at different phases. This section provides a generic description of various methods and their applicability to inventory of different pools for land-based projects and national greenhouse gas inventory. A list of methods for different carbon pools and their applicability is given in Table 9.1.

Table 9.1 Methodological options for estimating carbon pools (with details provided in the chapters)

Carbon pool	Methods	Suitability for carbon inventory of land-use systems
Above-ground biomass	Harvest method	- Not suitable, not often permitted, leads to disturbance of forest and even carbon emissions, expensive
	Carbon flux measurements	- Not suitable, expensive, requires skilled staff
	Satellite/remote sensing	- May not be suitable for multiple land-use systems and project activities - Not suitable for small projects - Practical methods still evolving
	Modelling	- Suitable for projections - Requires basic input parameters to be obtained using other methods
	Plotless method	- Suitable, but less suitable for periodic monitoring and dense vegetation
	Plot method	- Most suitable, cost-effective, commonly adopted and familiar
Below-ground biomass	Root extraction and weight measurement	- Expensive and not suitable - Requires uprooting of trees or grass and disturbs the soil
	Root to shoot ratio or conversion factor	- Most commonly adopted - Requires above-ground biomass estimate
	Biomass equations	- Requires input data on tree parameters, girth, height
Litter and deadwood	Litter trap	- Not always suitable in village or forest conditions, large effort needed
	Stock measurement	- Feasible, commonly adopted
Soil carbon	Diffuse reflectance spectroscopy	- Not suitable, expensive and requires skilled staff, has future potential
	Modelling	- Suitable for projections - Requires input data from other methods
	Field sampling and laboratory estimation	- Most suitable, commonly adopted and familiar method

Deadwood: standing deadwood biomass could be estimated using plot or plotless method. Fallen deadwood biomass could be estimated along with litter biomass.

9.3 Methods for Estimating Above-Ground Biomass

Harvest method The principle involves measuring the weight of tree and non-tree plant biomass in selected sample plots at a given point in time. It involves harvesting all the trees in the sample plots and measuring the weight of different components such as tree trunk, branches and leaves. Harvest method for non-tree biomass (shrubs, herbs, climbers and grass) also requires harvesting of non-woody biomass and of woody biomass, if any, from sample plots. The method gives the most accurate estimate of woody and non-woody biomass stock at the time of harvest and involves the following steps:

- Select the land-use category or project activity strata (see Section 10.3 for strata definition), sampling method and locate sample plots (based on methods described in Chapter 10).
- Harvest tree and non-tree biomass separately and, if necessary, separate the tree biomass into its components (trunk, branches and leaves) in each sample plot.
- Measure the fresh weight of tree and non-tree biomass and components and estimate the dry weight.
- Extrapolate dry biomass from the sample plot to per hectare values for tree and non-tree biomass.
- It may not be feasible to repeat the harvesting approach periodically because it destroys the vegetation in selected plots and new plots may have to be harvested each time. The harvest method is generally not regarded as suitable because it is expensive and destructive. Further, local land regulations and/or project design may not permit harvesting of tree or non-tree biomass. Finally, harvesting may lead to carbon emissions and loss of or disturbance to biodiversity. The method is not cost-effective, especially if large trees need to be harvested and weighed; however, it is suitable or can be adopted in following situations:
  - For estimation of non-tree annual biomass production such as grasses, herbs and even shrubs
  - For developing location- and species-specific allometric equations
  - For short-rotation commercial plantations, where plots are harvested every five to 10 years or so
  - For data on biomass stock or growth rates by component (trunk, branches, leaves)

Carbon flux measurement (Eddy covariance) method A range of methods exist for measuring and estimating the flux of CO_2 from vegetation cover over a land surface over different spatial scales. The method involves installing a chamber to enclose a small area or a particular component of an ecosystem (e.g. soil, stems, leaves). Changes in the concentration of CO_2 within the chamber or the difference between the concentrations in incoming and outgoing air are used to calculate the CO_2 flux. Methods exist for measuring the flux of CO_2 for an entire ecosystem (less than 1 km^2) without enclosures (Noble et al. 2000). The technique most commonly used is the "Eddy correlation/covariance technique", wherein measurements are continuous and semi-automatic (often at hourly intervals). The net flux of CO_2 entering or leaving the ecosystem integrated over an area typically of the order of 20 ha determines the overall net carbon exchange at a stand level.

Traditionally, the net ecosystem carbon exchange over multiple years has been estimated by quantifying temporal changes in biomass (Clark et al. 2001) and soil carbon (Amundson et al. 1998; Lal et al. 2001). Earlier, the method was employed to study CO_2 exchange of crops under ideal conditions during short field campaigns. The Eddy covariance method has emerged as an important tool for evaluating fluxes of CO_2 between terrestrial ecosystems and the atmosphere over the course of 1 year or more. This method is being applied in a nearly continuous mode to study CO_2 and water vapour exchange worldwide (Baldocchi 2003) and can be used to quantify how CO_2 exchange rates of whole ecosystems respond to environmental perturbations. The method provides an estimate of mass and energy exchange between vegetation surfaces and the atmosphere and allows direct and non-destructive measurement of net exchange of CO_2 comprising its uptake via photosynthesis and loss through respiration, evaporation and sensible heat.

The reasons for this method to emerge as an alternative way to assess ecosystem carbon exchange (Running et al. 1999; Canadell et al. 2000; Geider et al. 2001) include the following:

- It is a scale-appropriate method and can assess net CO_2 exchange of a whole ecosystem.
- It produces a direct measure of net CO_2 exchange across the canopy–atmosphere interface.
- The area sampled with this method, called the flux footprint, can extend from 100 m to several kilometres in length (Schmid 1994): in other words, it can spread over 10–100 ha, with near-continuous measurements possible.
- It can measure CO_2 exchange across a spectrum of timescales ranging from hours to years (Wofsy et al. 1993; Baldocchi et al. 2001).
- Data generated by this method provide key inputs to calibrate and validate canopy- and regional-scale carbon balance models.

The limitations of this method include the following:

- It is not feasible in areas that comprise different land-use systems, or landscapes with a mosaic of multiple land-use systems.
- It is expensive to establish.
- It is applicable only over flat terrains.
- It requires stable environmental conditions (wind, temperature, humidity and CO_2).

The technique is generally not regarded suitable for typical carbon mitigation or forest, plantation, grassland and cropland development projects because it is costly and needs highly trained staff.

Satellite or remote sensing method Remote sensing involves several techniques such as aerial photography, optical parameters and radar, which can be effectively used to track land-use changes in a project area. Further, remote sensing techniques provide an alternative to traditional methods for estimation, monitoring and verification of changes in areas under different land-use systems as well as in biomass production and growth rates (see Chapter 14 for more information on remote sensing). Remote sensing techniques provide spatially explicit information and enable repeated monitoring even in remote locations. The basic approach to applying remote sensing is to understand the relationship between the parameters of a forest stand (e.g. diameter at breast height (DBH), tree height, crown cover, basal area, and even biomass stock) and their spectral representation, depending on the characteristics of the study area and the sensor data used. Interpretation of remote sensing imagery therefore requires ground truthing and field measurements.

Remote sensing techniques for estimating biomass stocks are still evolving and are yet to be applied to land-based projects extensively; they are not currently suitable as the only method, but rather as a supplement to other methods, for land-based projects or for national greenhouse gas inventories because of limitations such as: (i) high cost, particularly for small-scale projects; (ii) technical and institutional capacity needed at the project level; and (iii) non-suitability to projects such as watersheds, village ecosystems or agroforestry, which involve a mosaic of small parcels (of a few hectares) of different land-use systems. Remote sensing techniques, when further developed with higher resolution, may become cost-effective for forest biomass inventory and national GHG inventory, particularly when the inventories have to be prepared periodically.

Modelling of carbon stock changes Models are available for projecting carbon stocks in biomass and growth rates of above-ground biomass of different commercial plantations as well as forest types. These models can be used to supplement field methods such as plot and plotless methods, where indicators of carbon stocks are measured and estimated. Models could be used to project changes in carbon stocks in biomass in forests and plantations. These growth models estimate biomass (kilograms/tree or tonnes/hectare) as a function of tree parameters such as DBH (in metres) and height (in metres). The biomass could be expressed in terms of volume (cubic metre) or weight (kilograms/tree). The volume could be converted to weight using the density of the tree species or woody biomass. The methods for developing biomass estimation equations are given in Chapter 17.

Such biomass estimation functions are normally available for specific tree species (also in Chapter 17) but not for many native or non-commercial tree species, stands of mixed species, natural forests, non-tree vegetation such as shrubs and grasslands or agroforestry systems. Separate equations are available for a whole tree and merchantable volume. The most popular ones estimate merchantable tree woody biomass. The application of models is limited because of the following shortcomings:

- Models of biomass stock or rate of growth may not be available for multiple land-use systems and project activities common to land-based projects.
- Equations developed for mature trees cannot be used for young trees and vice versa.
- Even when available, the models may not be applicable to local situations, as equations developed for one location may not always be appropriate to other

locations because of variation in varieties of a given tree species, vegetation and tree density.
- Developing the equations may require many trees of different sizes to be harvested and weighed and relating the weights to parameters such as DBH and height.

Despite these limitations, biomass equations, whenever available, provide the most suitable and sometimes the only approach to estimating biomass stock. Height and DBH, the parameters required for biomass equations, can be easily estimated from field methods. Such equations are not only a rapid method of estimating biomass stocks but also one that estimates standard error and the coefficient of determination.

Plotless method Plotless method involves measuring tree density and diameter (DBH) along a series of parallel sample lines (MacDicken 1997) and comprises the following steps:

- Select the land-use category or project activity strata, and locate sample plots (Chapter 10).
- Establish a series of parallel sample lines in each stratum.
- Locate sample points every 10 m along the sample line.
- At each sample point, divide the area into four quarters.
- Record the species name, DBH and height of the tree along with distance between the sample point and each tree or shrub.
- Ensure at least 100 measurements per stratum.

Using the data on the distance between the sample point and trees along the sample line, mean distance between trees in the plot can be estimated. Density of trees per hectare can then be calculated, either for each species or covering all the trees, using the estimated mean distance between trees. Above-ground biomass of trees can thus be calculated using DBH and height of trees and biomass equations.

The plotless method is more suitable for land-use systems with sparse tree density such as savannah and grasslands. This method is particularly useful for single period estimation, when large areas need to be covered in a short time and with limited personnel. The plotless method may not be suitable for periodic revisits and for measuring and monitoring changes in biomass or carbon stock of tree and non-tree vegetation in systems that involve multiple uses of land.

Plot method The principle of the plot method is to estimate the volume or weight of tree and non-tree biomass in a set of sample plots using the measured values of various indicator parameters such as DBH and height of tree. There are several variants of the plot method, namely quadrats (square or rectangular), circular plots and transects (long rectangular plots). The broad approach involves the following procedure:

- Select a land-use category or project activity; stratify and lay sample plots.
- Lay separate sample plots for trees, shrubs and ground-layer vegetation (herbs).
- Vary the size and number of sample plots depending on the type and size of the project and diversity of vegetation (which is discussed in Chapter 10).

- Record the species name, height and DBH for each tree or shrub.
- Estimate above-ground tree biomass per tree and per hectare using height and DBH data using different approaches, namely:
  - Biomass estimation equation
  - Harvest method within the plots
  - Calculating the volume of each tree using DBH, height and tree form data and then converting the volume to weight using wood density
- Estimate the biomass for non-tree vegetation such as shrubs and grasses by adopting the harvest method.

The plot method is the most commonly adopted method for assessing above-ground biomass of tree and non-tree vegetation. Application of the plot method is described in detail in Chapters 10–13. The merits of plot method include the following:

- Applicable to forests, plantations, grassland, shelterbelt and agroforestry systems
- Applicable equally to one-time measurement of biomass or long-term periodic monitoring through the "permanent plot" method
- Can be adopted by any team with minimal resources and technical capability
- Cost-effective
- Suitable for both sparse and dense vegetation
- Applicable to large or small patches of forest, plantation or grassland
- Suitable for both monoculture and diverse vegetation
- Suitable for both mature forests and young regenerating forests or plantations

9.4 Estimation of Below-Ground Biomass or Root Biomass

Below-ground or root biomass is necessary for natural forests, areas under natural regeneration, protected area and agroforestry systems. Root biomass is likely to be important for afforestation, reforestation, watershed and grassland reclamation projects. Methods for estimating below-ground tree biomass include the following:

(i) Root extraction and weight measurement
(ii) Default root to shoot ratio
(iii) Biomass equations

Root extraction and weight measurement method The method of root extraction and weight measurement involves measuring the quantity of root biomass present in a given volume of soil extracted from a known depth, which is normally 30 cm because most fine roots are confined to this shallow depth. The method involves the following steps (MacDicken 1997):

- Using a core sampler to remove a known volume of soil from a selected depth
- Washing the core soil sample and extracting the roots by separating the soil
- Measuring the fresh and dry weight of the root biomass for the selected core volume
- Extrapolating the root biomass from the core volume to unit area, such as 1 ha, for a given depth, say 30 cm

Measuring root biomass is complex, time consuming and expensive (Cairns et al. 1997). Root biomass is not measured in most land-based projects, and alternative methods such as default root to shoot ratio and biomass equations are adopted. The method is described further in Chapter 11.

Default root to shoot ratio Root biomass is normally within a small range of proportion of above-ground biomass. A review by Cairns et al. (1997) covering more than 160 studies from tropical, temperate and boreal forests estimated a mean root to shoot ratio of 0.26 with a range of 0.18–0.3. Thus, it may be practical to use a mean default value of 0.26 for estimating the root biomass in most forestry projects.

Biomass equations Regression equations have been developed linking root biomass to above-ground biomass. Cairns et al. (1997) have developed a set of equations for tropical, temperate and boreal forest types. Refer to the equations given in Chapter 11. These equations provide reliable estimates with high coefficient of determination.

Root biomass of non-tree vegetation in land-use systems such as grassland, cropland and savannah can be estimated by measurement. The broad method is identical to that of root extraction using a core sampler, described for tree roots. The detailed steps are described in Chapter 11. It is very important to make an expert judgment as to when root biomass measurement is required. Root biomass is normally estimated if grassland or degraded forest land has been converted to cropland or even to managed grassland involving disturbance to topsoil (Chapter 4).

9.5 Estimation of Litter and Deadwood Biomass

Litter and deadwood biomass could be estimated using the two methods, namely (i) production measurement and (ii) stock change. Deadwood consists of standing and fallen deadwood.

Annual litter and fallen deadwood production (tonnes/hectare/year) Estimating annual litter and fallen deadwood production is a very complex and lengthy procedure, involving the following steps:

- Selection of the land-use category or project activity, stratification and location of sample points
- Installation of a large number of rectangular or circular litter traps on the floor of forests or plantations and protecting them from damage for many years
- Monthly collection and weighing of litter collected in the litter traps in sample plots and estimation of dry weights
- Extrapolation and computation of annual litter production per hectare

Default values are available from literature for some forest and plantation types. Default values could be used since litter accounts for a small fraction of total biomass (<10%) and it is too complex to install and maintain the litter traps, apart from the cost and technical effort involved.

Stock change method Litter and fallen deadwood can be estimated by measuring the stock in sample plots at two points in time and calculating the difference. The sample plots selected for tree or shrub measurements can be used for the estimation, adopting the following procedure:

- Select the sample plots used for tree and shrub biomass estimation.
- Collect and weigh all the fallen litter and deadwood from the sample plots.
- Estimate the dry weight of litter and deadwood.
- Repeat the measurement at two time periods.
- Estimate the difference between the two measurements.
- Extrapolate litter and deadwood stock from sample plots to per hectare values.

Biomass of standing deadwood could be estimated using the plot method described in Section 9.3. The procedure described for above-ground tree biomass in Section 9.3 (details in Chapter 11) is applicable to estimating biomass of standing deadwood and involves measuring DBH and height of standing dead trees from plots or quadrats selected for tree or shrub measurement. Estimation of standing deadwood requires very little additional effort, since it can be measured along with the measurements of tree and shrub biomass.

9.6 Estimation of Soil Organic Carbon

Soil organic carbon is a critical carbon pool for majority of land-use categories and afforestation, reforestation, land reclamation, grassland management, shelterbelt and agroforestry projects. Soil carbon pool is particularly important for projects involving savannah, cropland, grassland and rangeland. Soil carbon stock is the highest in the upper soil profile (0–15 cm), which should be sampled most intensively (Richter et al. 1999). Soil organic carbon is routinely estimated for all forestry, grassland and cropland conservation and development projects by any of the following methods, namely diffuse reflectance spectroscopy, modelling and wet digestion or titrimetric determination.

Diffuse reflectance spectroscopy Diffuse reflectance spectroscopy (DRS) is a technology for characterization of the composition of materials based on the interaction of visible–infrared light (electromagnetic energy) with matter. Soil samples are illuminated with an artificial light source and the reflected light diffusing from the sample is measured. Reflectance readings are taken with a spectroradiometer, and each wavelength band is expressed relative to the average of the reference readings for each property of soil. The method has the potential to increase efficiencies and reduce costs in large-area applications (soil survey, watershed management and soil quality indicators).

The main merits of DRS are its repeatability and speed compared to conventional soil analyses; a single operator can comfortably scan several hundred samples a day. The repeatability among laboratories is expected to be greater with DRS than with conventional methods of analysis. The speed of analysis and the ability to estimate many soil properties from a single robust measurement are major advantages of DRS, especially for analysing soil properties that are time consuming to measure by conventional methods.

The main limitations of DRS are the need to build calibration libraries for a given population of soils for all the soil properties of interest and the complexity of the data analysis this involves. The method requires specialized equipment, a well-equipped laboratory and trained staff. The technology could be increasingly used in a wide range of soil studies and surveys, and spectrometers are likely to become standard equipment in soil laboratories.

Modelling of soil carbon Soil carbon dynamic models include CENTURY (CENTURY 1992) and RothC (Coleman and Jenkinson 1995).

CENTURY simulates long-term dynamics of carbon, nitrogen and phosphorus for different plant soil systems. The model simulates the flow of carbon, nitrogen, and phosphorus through the plant litter and different inorganic and organic pools in the soil. The advantage is that the model can be applied to forest, grassland, savannah, and cropping systems or projects at the plot, project, regional and national level. The limitations include data needed such as precipitation, maximum and minimum air temperature, lignin, nitrogen, phosphorus and sulphur in the plant material, texture of the soil, initial contents of total carbon, nitrogen and phosphorus in the soil, amounts of agricultural inputs used during the management cycle, etc.

RothC computes changes in organic carbon content in tonnes/hectare. RothC requires input of a large number of variables such as weather data (monthly mean temperature, total monthly precipitation and open pan evaporation), soil data (percentage of clay in soil and soil depth), land management data related to carbon simulation and monthly inputs of organic matter to soil. The limitation of the model is that the data needed for diverse land-use systems are hard to come by.

Simple regression models for predicting soil carbon accumulation rates or stock changes have not evolved as much as those for, say, above-ground biomass. Such models are specific to soil, vegetation or species and will have limited application.

Wet digestion or titrimetric determination (Walkley and Black method) The method of wet digestion or titrimetric determination involves rapid titration procedure for the estimation of organic carbon content of the soil (Kalara and Maynard 1991). Organic matter is oxidized with a mixture of $K_2Cr_2O_7$ and H_2SO_4. Unused $K_2Cr_2O_7$ is back-titrated with $FeSO_4$. The dilution heat of concentrated H_2SO_4 with $K_2Cr_2O_7$ is the sole source of heat. The soil is digested by the heat of dilution of H_2SO_4 and thus organic carbon in the soil is oxidized to CO_2. Among the various methods, wet digestion (Walkley and Black) is the most commonly adopted and cost-effective method, which involves the following procedure:

- Selection of the land-use category or project activity strata, sampling method and location of sample plots
- Collection of soil samples at two depths (0–15 cm and 15–30 cm) from each stratum
- Estimation of bulk density
- Estimation of organic matter or carbon content in the soil sample in the laboratory using the Walkley and Black method
- Calculation of carbon stock in tonnes of carbon/hectare using organic matter content, bulk density and depth of soil

9.7 Conclusions

This chapter provided an overview of different methods available for estimating the stocks and growth rates of different carbon pools. Carbon inventory of different carbon pools requires adoption of different field, laboratory and modelling techniques. Multiple methods are available for different pools. The selection of a suitable method for a given pool depends on the land-use category or project type, size of the project area, accuracy needed, cost-effectiveness, infrastructure and technical capacity available. The details of these methods are described in later chapters. The focus of this handbook in the remaining chapters is on adopting the carbon "Stock-Difference" method using "permanent plot" technique for carbon inventory.

Chapter 10
Methods for Estimating Above-Ground Biomass

Above-ground biomass includes all biomass in living vegetation, both woody and herbaceous, above the soil including stems, stumps, branches, bark, seeds and foliage. Above-ground biomass is the most visible of all the carbon pools, and changes in it are an important indicator of change or of the impact of an intervention on benefits related to both carbon mitigation and other matters. Above-ground biomass is a key pool for most land-based projects. The features and the need for measuring and monitoring above-ground biomass, its importance to national greenhouse gas inventory and different project types as well as the frequency of measurement of the pool are described in Chapter 4. The different methods available for estimation and monitoring of above-ground biomass pool are described in Chapter 9. Among all the methods described in Chapter 9, the "plot method" is described in detail in this chapter. The rationale for selecting the "plot method" as the most suitable method includes the following factors:

- Applicability to baseline as well as project scenario measurements
- Applicability at project development and project monitoring phases
- Applicability to national greenhouse gas inventory estimation
- Suitability for all project types and projects of different tree sizes (mature and young) and density (dense and spare) of tree vegetation
- Simple and cost-effective
- Suitability for long-term monitoring

The "plot method" is extensively used in forest inventory programmes and by project managers and evaluators for estimating and monitoring roundwood production or carbon stock changes. Researchers in forestry, ecology and agriculture routinely adopt this method not only for estimating the changes in biomass stock, but also for monitoring biodiversity and production of commercial timber, fuelwood and grass. The "plot method" is further used for estimating biomass changes in cropland as well as grassland projects. The "plot method" is described in reports, manuals and books Special Report of Intergovernmental Panel on Climate Change on Land Use Land-Use Change and Forestry (Watson et al. 2000), Winrock Carbon Monitoring Guideline (MacDicken 1997), FAO (Brown 1997), Revised IPCC 1996 Guidelines (IPCC 1996), IPCC Good Practice Guidance (IPCC 2003), USEPA and LBNL (Vine and Sathaye 1999), CIFOR Methods (Hairiah et al. 2001), GHG Inventory

Guidelines 2006 (IPCC 2006) and Forest Inventory (Kangas and Maltamo 2006). This chapter presents a detailed description of methods, procedures, and steps for measurement, estimation and monitoring of above-ground biomass stocks and changes in those stocks.

Broad approaches to measurement, estimation and monitoring of carbon stocks, namely carbon "Gain–Loss" and "Stock-Difference" are presented in Chapter 9; the plot method can be applied to both these approaches.

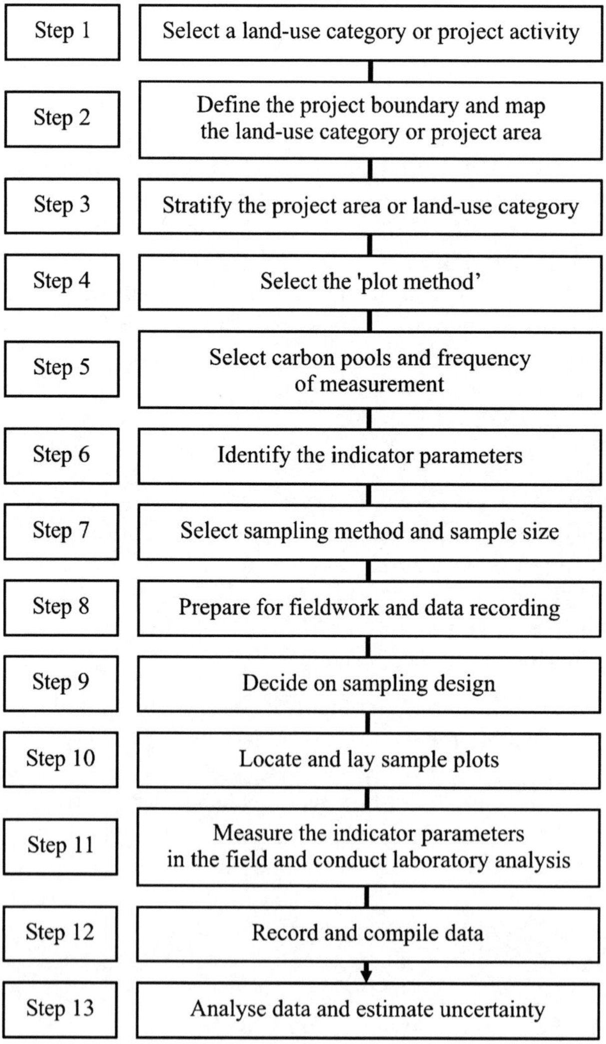

Fig. 10.1 Steps in measurement and estimation of above-ground biomass stock

The methods and steps are described for projects related to carbon mitigation as well as roundwood production. The methods and steps are also applicable to national greenhouse gas inventory for land-use categories such as forest land, cropland and grassland.

Procedure for measurement and monitoring of above-ground biomass stock All programmes and projects require estimation of above-ground biomass stock at a given point in time, annual growth rates as well as changes in stocks over a period of time. The broad steps for estimating the stock of above-ground biomass at a given point in time are given in Fig. 10.1.

10.1 Selection of Land-Use Category, Project Activity or Vegetation Type

Every project will have a set of activities aimed at achieving the project goals. A project may have a single activity such as planting eucalyptus species with a single set of management practices. Some projects may have multiple activities, where a part of the project area is under monoculture plantation and the rest could be natural regeneration of degraded forest land. These activities will have different above-ground biomass accumulation rates. Further, the density, rotation period and silvicultural practices such as irrigation and fertilizer application may vary. These different activities and management systems will have implications for carbon inventory and decisions on plot sizes, frequency of monitoring and parameters for measurement. The selection could be along the following lines:

- Land-use category (forest land, grassland and cropland), subcategory (based on soil or topography), vegetation type (evergreen and deciduous forests)
- Project activity (afforestation, avoided deforestation and grassland reclamation) and management system (density of plantation, rotation period, fertilizer application or irrigation)

10.2 Definition of the Project Boundary and Mapping of the Land-Use Category or Project Area

It is important to define the project boundary and prepare a map of the project area, demarcating the areas under different activities and management systems for carbon inventory. The project boundary definition and methods for defining project boundaries are given in Chapter 8. The project boundary may incorporate all the land-use categories, project activities and management systems. The project boundary and the extent of area under each of the land-use systems and activities should be spatially represented on a map. The following procedure could be adopted for defining the boundary and mapping.

(i) *Select the land-use category and project activities* A project may include multiple land-use categories, project activities and management systems. All the land-use categories, project activities and management systems should be selected separately for carbon inventory, for example:

- Land-use categories such as forest land, grassland and cropland
- Multiple activities such as promotion of natural regeneration on degraded forest land, monoculture plantation on grasslands and agroforestry in cropland
- Plantation activity may include short-rotation as well as long-rotation species
- Plantation activity may have high and low density plots and could be with or without fertilizer application

(ii) *Estimation of area under the project or project activity* Data on area under different activities are required for sampling as well as for estimating the above-ground biomass stock and changes. Project area estimation is required at the following phases of a project:

- *Project development phase* Obtain the area under each land-use category and project activity proposed in the project document.
- *Project monitoring phase* Obtain the actual area under each of the activities from the project authorities. The area actually brought under the project activity may differ from the area proposed at the project development phase. If the project activity is implemented in phases, obtain the data annually.

(iii) *Map preparation* Data on the total project area and area under different activities, subactivities and management systems should be obtained and spatially marked on a georeferenced map with a grid. The procedures for map preparation and boundary marking are described in Chapter 8. Map preparation involves the following key steps:

- *Historical land records* Collect any historical land-use records over the last 10–20 years to understand the trends in land-use change and to project the future land-use pattern under the baseline scenario.
- *Maps* Collect all the maps available for the project location: soil, vegetation, land tenure and land use.
  - If maps from satellite imagery and aerial photography are available, it would be very useful to collect them.
  - Using different maps, select the most important map showing the features relevant to the project, for example, land use or soils.
- *Overlay* Overlay different maps with various features on a geo-referenced map with many identifiable landmarks on the ground and mark the existing land-use systems such as cropland, grassland, water bodies or settlements.
- *Boundary* Mark the boundary of different project activities and management systems on the georeferenced grid map with relevant features such as soil quality and land use.
- *GPS readings Mark the GPS readings of the polygons (plots of different shapes*) under different activities and of different parcels of land.

- *GIS maps* If GIS facility is available, which is becoming increasingly common, different maps with various land features as well as the project activities could be overlaid along with positions needed for boundary and areas. These spatially oriented maps on a GIS platform would help the project managers and monitoring teams to
  - Understand the land-use changes
  - Map the area brought under different activities over the years
  - Track the area brought under different management activities such as area harvested periodically or treated differently
  - Locate and sample the plots and revisit the "permanent plots" periodically
  - Mark areas for estimating leakage
  - Store existing data
  - Easily record changes throughout the project period
- *Updating of maps* It is necessary to periodically update the maps, depending on any new information on changes in land use, such as implementation of project activities or area subjected to harvest.
- *Maps from remote sensing* Data on land-use changes, particularly the historical ones, can be obtained from interpreting satellite images over the area. Periodic updates of remote sensing maps will help in assessing
  - Changes in boundaries of different land-use systems
  - Changes in crown cover
  - Rate of implementation of project activities
  - Biomass stocks and changes

10.3 Stratification of the Project Area or Land-Use Category

Stratification is disaggregation of land area into homogeneous subunits. Land area of any project will consist of strata with varying physical and biological features, subjected to different management practices, and carbon stocks may vary because of these and other features. Stratification helps in obtaining a better representation of the land or project activities while sampling reflects the diversity of conditions that contribute to carbon stocks and minimizes costs. Stratification reduces sampling error and sampling effort by aggregating those spatial components that are homogeneous. Stratification of land area is required for the baseline as well as the project scenario (Fig. 10.2), which may be the same or different. Multistage stratification may be required, for example, to highlight broad land-use categories, features of the land, the main project activity and management systems within a project activity (as described in Fig. 10.2). The strata for sampling and monitoring refer to the last stage of disaggregated homogeneous land area or project activity such as (i) high-density short-rotation plantation on degraded forest land and (ii) grassland development with irrigation. In this handbook, a stratum refers to the last stage homogeneous unit of multistage strata for a given land-use category or project activity.

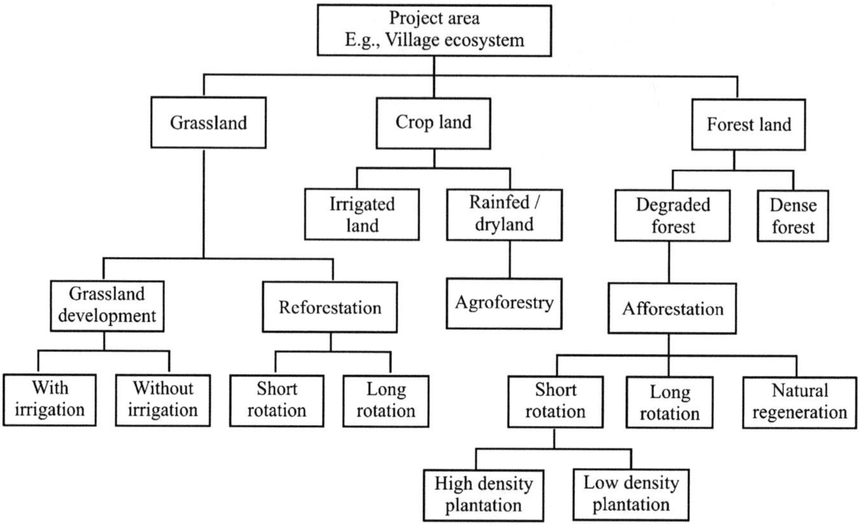

Fig. 10.2 Stratification for a multicomponent project in a village ecosystem

10.3.1 Stratification for Baseline Scenario

The baseline scenario requires estimates of carbon stock at the beginning of the project, before project activities are implemented. Further, carbon stock changes in the control plots need to be monitored to estimate the changes under the adjustable or moving baseline scenario, where carbon stocks are expected to change over the years (Chapter 6). The land area considered for the project for the baseline scenario may consist of diverse land conditions in terms of

- *Pre-project land-use category* degraded forest land, grassland, cropland
- *Soil quality* soil erosion status and topography
- *Extent of dependence or use of the land resource* grazing, fuelwood collection, proximity to the settlements
- *Tenure/ownership* state-owned, farmland, community land
- *Current management system* regulated grazing or open access system, irrigated, rainfed or fallow cropland

10.3.2 Stratification for Project Scenario

Carbon inventory during the project scenario requires estimation of stocks and changes in stocks annually or any other selected frequency of monitoring above-ground biomass. Carbon stocks may have to be estimated separately for the total area under each activity (e.g. short-rotation plantation or natural regeneration) and for the total project area consisting of multiple activities (e.g. area under short rotation *and* that under natural regeneration and so on). Therefore, each of the activities implemented needs to be sampled. Stratification in the project scenario may depend on the following factors:

- *Preliminary status of the land* Status of land at the beginning of the project, that is the baseline scenario:
    - Pre-project land-use system, soil quality, extent of dependence on the land resource, tenurial status and current management system
- *Project activity* Short- or long-rotation plantation, natural regeneration, agroforestry and grassland management
- *Species* Monoculture, multispecies plantation and natural regeneration
- *Management system* Density of plantation, irrigation and fertilizer application

Stratification during the project scenario could be identical to that adopted for the baseline scenario or different from it to reflect the impacts of project activities implemented on the baseline strata.

(a) *Disaggregation of baseline strata* If a baseline stratum, for example, a land-use system, is subjected to multiple project activities in the project scenario, disaggregation of that stratum may be necessary.

- For example, if part of the grassland stratum of baseline is under short-rotation species and the rest under natural regeneration activity, disaggregation of the baseline stratum may be required to represent the two project activities that have different implications for carbon inventory.

(b) *Aggregation of baseline strata* If multiple baseline strata are covered with a single project activity and if the soil or other parameters characterizing the strata are identical, the multiple strata could be aggregated.

- For example, if grassland and degraded forest land strata of the baseline scenario with identical initial characteristics (e.g. vegetation, soil quality or slope) are planted with a monoculture short-rotation plantation with similar density and management system, the two strata could be aggregated.

10.3.3 Stratification Under Land-Use Change

The project scenario may involve land-use change from the baseline or the land use may remain unchanged, with land remaining in the same category as that in the baseline, but with improved managed system, which will have implications for carbon inventory and methods to be adopted for carbon stock estimation.

- *Land remaining in the same category* The project scenario will involve no change in the land-use compared to the baseline; however, the project activity may improve land management or prevent the expected change in land use.

   - *Avoided deforestation* where forest land of the baseline scenario would remain as forest land in the project scenario.
   - *Improved grassland management* the land use and the land cover may remain the same in the project scenario as in the baseline, but with improved management system.

- *Land converted to another category* The project scenario will involve change in land use or land cover in the project scenario as a result of project activities.

   - *Afforestation* Grassland converted to short-rotation forest plantations in the project scenario.
   - *Agroforestry* Cropland converted to agroforestry.

10.3.4 Approach to and Steps in Stratification

A stratification procedure for the baseline as well as the project scenario, which requires identical steps, is presented in this section:

Step 1: Define the project boundary as described in Section 10.3.

Step 2: Obtain maps of the project area and overlay the different maps representing, for example, land-use systems, soil and topography under the baseline scenario.

Step 3: Overlay the project activities on land-use systems in the baseline scenario, such as degraded forest land or grassland.

Step 4: Identify the key differentiating features for stratification of land-use systems in the baseline scenario that are likely to impact carbon stocks:

   - *Current land-use* such as open access grazing, controlled grazing, fuelwood extraction and rainfed cropping
   - *Soil quality* good, moderate and low
   - *Topography* levelled, sloping, hilly terrain

Step 5: Collect all the information available from secondary sources as well as from participatory rural appraisal (Chapter 8).

Step 6: Stratify the area under the baseline scenario:

   - Delineate areas under different project activities.
   - Overlaying the delineated areas with key features of land-use systems that are critical for estimating baseline carbon stocks.
   - Mark the strata to be brought under different project activities spatially on the project map.

Step 7: Stratify the area under the project scenario:

   - Locate the project activities on the baseline scenario strata spatially.
   - Mark spatially the different strata representing different project activities, land-use systems and other features; however, each stratum is homogeneous within itself.

10.3.5 Application of Remote Sensing and GIS for Stratification

A variety of remote sensing data are available, which can be used for defining project area or land-use category as described in Chapter 8. The data products can be in the form aerial photography or satellite imagery. This means that different data sets may cover different time series, a point significant for developing a baseline for above-ground carbon. More information on remote sensing and its use in carbon inventory is given in Chapter 14. Several important criteria govern the selection of remote sensing data and data products for defining the project area or land-use category (IPCC 2006).

- *Adequate categorization of land use* The data should help in distinguishing between different types of land use; for example, grassland and forest land are seldom a problem to monitor, but natural forest and degraded forest can be difficult, depending on the remote sensing product.
- *Appropriate spatial resolution* The data should be available at an appropriate resolution, since the spatial resolution determines how well the data can be classified into different levels of land-use classes. For most project-based assessments, a fine resolution (25 m or lower) is needed to categorize different land uses.
- *Appropriate temporal resolution* Because land use may change with time, the data should offer adequate temporal resolution as well for estimating land-use conversions. To be able to assess the process of change, data need to be analysed across a given time span. Depending on geographical location and vegetation, it is also important to take seasonality of vegetation into account. Usually, peak vegetation periods are the ones most easily used.
- *Ground-truthing* To validate the data from remote sensing, they need to be compared against data obtained by other means. Interpretation of remote sensing data is cross-checked with empirical data on vegetation in the land-use systems to be assessed. For above-ground biomass, actual biomass inventory data are very useful but management plans, land-use maps or data from participatory rural appraisal can also be used.

Geographical information system (GIS) can be used not only for interpreting actual remote sensing data but also for synthesizing the data collected over the project area. Further, it is a perfect system for storing and adding data over time. Several user-friendly programs are available to be used on regular PCs.

10.4 Selection of a Method for Estimation of Above-Ground Biomass: the "Plot Method"

Several methods are presented for estimating above-ground biomass in Chapter 9, of which the "plot method" is the one most commonly used. The method, versatile, cost-effective and applicable to baseline as well as project scenario, is described in detail here. The "plot method" is used in preparing a forest inventory and estimating biomass in grassland, crop productivity and timber and fuelwood production. "Plot method" is also among the methodologies approved by the Clean Development Mechanism for afforestation and reforestation projects (http://cdm.unfccc.int) under the Kyoto Protocol. The method involves selecting plots of an appropriate size and number, laying them randomly in the selected strata, measuring the indicator parameters (e.g. tree DBH, height or grass production), using different approaches such as allometric functions to calculate the biomass and extrapolating the value to per hectare and for the total project area. These sample plots could also be used for assessment of biodiversity, land degradation and soil fertility improvements.

10.5 Selection of Appropriate Frequency of Measurement for the Above-Ground Biomass Pool

The frequency of measurement and monitoring of above-ground biomass pool depends on the land-use system, soil quality, species and management systems (refer to Chapter 4 for details). The frequency is different for the baseline and for the project scenario and also depends on the biomass stock and its rate of growth. Frequency of monitoring has implications for carbon inventory due to the effort and cost involved:

- *Baseline scenario* The frequency will depend on the rate of above-ground biomass growth, which is likely to be low for most baseline scenario situations. Thus, the above-ground biomass could be monitored once in 3–5 years. The frequency of monitoring for avoided deforestation projects, with high biomass stock under baseline scenario, could also be 3–5 years.
- *Project scenario* The frequency will be determined by the type of project activity and the rate of growth of above-ground biomass. Fast growing species, such as those grown intensively for bio-energy plantations, may require frequent monitoring.

It is important to decide on the frequency of monitoring above-ground biomass stocks so that resources can be planned for and allocated accordingly.

10.6 Identification of the Parameters to be Measured for Estimating the Above-Ground Biomass Pool

The goal of measurement and monitoring is to estimate the stocks of above-ground biomass or its rate of growth on per hectare basis as well as for the total project area. This requires identification and selection of a key set of indicator parameters. The parameters to be selected depend on the method adopted; those required for the "plot method" are presented here. The most commonly used parameters are as follows:

(i) *Name of the species* The first parameter to be recorded is the plant form, namely tree, shrub, herb or liana, followed by the name of the species. Among trees, species differ in shape, size, rate of growth and wood density. It is also important to estimate the density of trees (number per unit area) of each species in the sample plots and per hectare. Names of species are important even for non-tree plant forms such as shrubs, herbs and grass. Biomass for tree species is estimated as volume or weight per tree, which can be extrapolated to per hectare based on the density and distribution of each species. While recording the species name and number, it is desirable to record other features such as:

- *Regeneration* naturally regenerated or planted
- *Status of tree crown* percent damaged or full crown
- *Status of the tree* living, dead and standing, or dead and fallen

(ii) *Diameter or girth at breast height for trees* Size, usually measured in terms of diameter or girth at breast height (DBH or GBH), is one of the most important parameters and represents the volume or weight of a tree, which can be converted to biomass per unit area (tonnes/hectare or tonnes/hectare/year). The diameter and height can be used for estimating the volume by simple equations; DBH values can also be used in allometric functions to estimate volume or biomass per tree or per hectare. Usually, DBH is easy to measure in the field and, by appropriate marking, the measurements can be repeated over time. The breast height in DBH is normally taken to be 130 cm above the ground. The measurement techniques are described later in the chapter (Section 10.11).

(iii) *Height of trees* Next to DBH, height is the most important indicator of the volume or weight of a tree and used in many allometric functions along with DBH. Measuring the height of tall trees, especially those with overlapping canopies, requires instruments and may introduce errors.

(iv) *Indicator parameters for non-tree species* Height and DBH are not measured for non-tree species such as herbs and grasses; biomass is estimated in terms of weight per unit area by actually harvesting and weighing all the herbs and grasses in the sample plots.

The parameters to be monitored for estimating above-ground biomass are listed in Table 10.1.

Table 10.1 Parameters for estimation

Carbon pool	Parameters to be recorded
Above-ground biomass of trees and shrubs	Name of the species DBH (cm) Height (m) Origin: regenerated or planted Extent of crown: full crown or percent crown damaged Status: dead or living
Above-ground biomass of herb or ground-layer vegetation	Name of the species Density (number/plot) Fresh weight of herb layer biomass (g/m^2) Dry weight of herb layer biomass (g/m^2)

10.7 Selection of Sampling Method and Sample Size

Sampling includes deciding on the number, size and shape of the plots, a step often ignored by project developers and managers because of the perceived complexity of the methods. Two approaches can be considered for measurement and estimation of carbon in land-use systems, namely complete enumeration and sampling. Complete enumeration, measurement and monitoring of all trees and non-tree plants in different land-use systems is time-consuming, very expensive and not even necessary to get a reliable estimate of biomass. A carbon inventory based on appropriate sampling can yield reliable estimates at a limited cost and human effort. Thus, the main goal of sampling is to get a reliable estimate with minimal cost. Sampling methods include simple random sampling, stratified random sampling and systematic sampling. This section presents the principles of sampling, the accuracy and precision needed, the methods for choosing sample size and shape of the plots and practical steps to be followed. Sampling is crucial to measuring and monitoring carbon stock changes. Several books (Johnson et al. 2000; Shiver and Borders 1996; De Vries 1986; Wenger 1984) and guidebooks (MacDicken 1997; Pearson et al. 2005b; FAO 2005; http://cdm.unfccc.int) are available to assist sampling, including IPCC Good Practice Guidance (IPCC 2003).

10.7.1 Sampling Principles

Sampling enables conclusions to be drawn about an entire population by observing only a portion of it. Sampling theory provides the means for scaling up information from the sample plots to the whole project area or even to a regional and national level (IPCC 2003). Thus, measurements of indicators of carbon stocks made on a small set of sample plots can be extrapolated to per hectare, for the strata and the whole project area or the land-use category. Field sampling is needed for all methods of carbon inventory – plot method, harvest method and even remote sensing techniques – that require ground-based data from sample sites for interpretation and verification. Standard sampling theory relies on random selection of a sample from the population so that each unit of the population has an equal probability of being included in the sample.

Accuracy and precision Sampling involves two common statistical concepts, namely accuracy and precision (IPCC 2003; Pearson et al. 2005b). Accuracy is a measure of how close the sample measurements are to actual values. Inaccurate or biased measurements will move the average away from the actual value. Precision is a measure of how well a value is defined. In the case of carbon inventory, precision shows how closely the results from different sampling points or plots are grouped. Figure 10.3 shows:

(a) Points are close to the centre and therefore accurate but widely spaced and thus not precise.

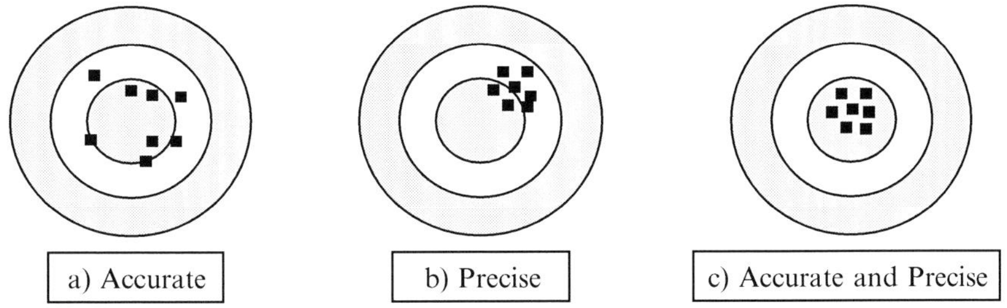

Fig. 10.3 Accuracy and precision

(b) Points are closely grouped and therefore precise, but far from the centre and thus not accurate.
(c) Points are close to the centre as well as tightly grouped, and thus both accurate and precise.

Accuracy and precision reflect how well the measurements estimate the true value of tree variables such as diameter, height and area covered by a stand of trees. An unbiased estimate will depend on repeated measurements being similar (precise) and averaging close to the true value (accuracy). A carbon inventory requires measurements that are both accurate, or close to the population values, and precise or closely grouped. Sampling involves selecting plots that are appropriate in size, number and location, which contributes to accuracy and precision. The level of precision required for carbon inventory has direct implications for inventory costs. The level of precision should be determined at the beginning of a project, and could vary from ±5% to ±20% of the mean. The lower the precision, the lower the confidence that the change in carbon stocks over time is real and due to project activity. The chosen level of precision will determine sample sizes for each project activity (Pearson et al. 2005b).

Confidence interval The representativeness of the estimate, or precision, is indicated by the confidence interval. Normally a 95% confidence interval is used, which implies that 95 times out of 100, the estimated value lies within the limits of twice the standard deviation.

10.7.2 Type and Shape of Sample Plots

10.7.2.1 Type of Plots

Two types of sample plots could be adopted for land-based projects, namely permanent sample plots and temporary sample plots: the type of vegetation determines which of the two is to be adopted.

Permanent plots are used mainly for measuring changes in carbon stocks in perennial vegetation where, for example, the trees may have to be measured over a number of years. This approach is suitable for most of the land-based projects involving tree carbon pools:

- Forests and plantations
- Agroforestry
- Shelterbelts

Permanent sample plots are generally regarded as statistically more efficient in estimating changes in forest carbon stocks compared to temporary plots because typically there is high covariance between observations taken at successive sampling events in temporary plots (Avery and Burkhart 1983). The disadvantage of permanent plots is that their location can be known and they could be treated differently (e.g. application of fertilizer and irrigation to increase the rate of carbon stock accumulation). Further, these plots might be damaged or destroyed by fire or other disturbances during the project period. This disadvantage could be overcome by ensuring that silvicultural practices are identical for the whole area under a given activity during the period of monitoring and verification: if the plot is damaged, say because of fire, new sample plots can be chosen, with identical soil and plant growth patterns.

Temporary plots The location of temporary plots could vary from year to year or over a number of years. In temporary plots, the measurements are made for a given year and biomass is calculated only for that year. Next year, biomass is estimated from a different plot. Such approach is suitable for projects involving annual vegetation:

- Estimation of grass production in grassland reclamation and savannah projects
- Estimation of production of herbaceous vegetation in forests and plantations.

The advantages of temporary plots are that they may be established more cost-effectively to estimate the carbon stocks of relevant pools and that the sampling would not be affected by disturbances. The main disadvantage of the temporary plots is related to precision in estimating change in forest carbon stocks (IPCC 2003). In temporary plot approach, individual trees are not tracked (Clark et al. 2001) and the covariance cannot be estimated, which makes it difficult to attain the targeted precision level without measuring a large number of plots. Thus, the cost advantage of using temporary plots may be lost by the need to establish more temporary plots to achieve the targeted precision.

Thus, the permanent plot approach could be adopted for forests, plantations, agroforestry and other perennial vegetation systems whereas the temporary plot approach could be adopted for annual vegetation systems such as grassland and cropland.

10.7.2.2 Shape of Plots

The shape of a plot has implications for accuracy and ease of measurement. The standard plot shapes used in vegetation studies are rectangles and squares, although strips and circles are also used.

Rectangular or square plots Establishing rectangular or square plots involves measuring out the length or the breadth and using the diagonal to ensure a true right angle at each of the four corners. A square or a rectangle is the most commonly adopted shape for plots for estimating biomass in most vegetation types including forests, plantations, agroforestry, shelterbelt, grassland and cropland because of the following factors:

- Easy to lay out and ensure square corners
- Suitable for young and mature forests as well as for non-tree vegetation such as grassland and cropland
- Suitable for large plots (e.g. 50 × 50 m or 100 × 100 m) or small plots (e.g. 1 × 1 m or 5 × 5 m)
- Easy to establish corner points for drawing the boundary for periodical visits and long-term monitoring
- Easy to record GPS readings and to locate the sample plots in later years for monitoring.

Circular plot Circular plots could be adopted for vegetation types such as trees, herbs and grasses. It is easier to draw circular plots of small dimension in the field. However, a circular plot is not popular because of the following reasons:

- Difficult to mark a circular plot in forests and plantations where large trees exist
- Difficult to verify the boundary line and area of the plot
- Difficult to mark the boundary line for periodic visits and not suitable for long-term monitoring
- May not be suitable for agroforestry, shelterbelt and avenue trees
- Not very efficient (Loetsch et al. 1973), since, as the perimeter increases, so will the numbers of trees on the edge of the plot

Strip plots Strip plots or belt transects are long and narrow rectangular plots, normally used for studying rare populations. A strip plot is not usually adopted for biomass studies due to the following reasons:

- Difficult to mark long strips in the field, especially if there are large trees.
- Size of the project area may be a limitation. For example, a sample plot of 2,000 m^2 requires a strip 200 m long, and several such sample strips may be required.
- Difficult to draw boundary lines for the long narrow strips for periodic monitoring.

10.7.3 Number of Plots

Deciding on the number of plots to be measured or monitored is a critical step in assessing carbon stock changes in a land-use category or project. There is a trade-off between precision and cost. The more spatially variable the carbon stocks in a project, the more sampling plots needed to attain a given level of precision at the same confidence level (IPCC 2003). This has cost implications for planning and implementing a monitoring plan. The number of sample plots should be chosen with statistical rigour to get a correct assessment of the impact of a land-based project on carbon stocks, roundwood production or soil organic matter. The number depends on the desired precision, size of the project, variation in vegetation parameters, available budget and cost of monitoring. Sampling methods and procedures can be complex, forcing many researchers to adopt a standard number of plots based on the visible heterogeneity of vegetation, soil and other conditions. There are a number of statistical approaches to and formulae for determining sample size (Usher 1991; Kangas and Maltamo 2006). A commonly adopted approach is illustrated here. The following steps and calculations are necessary for arriving at the number of sample plots:

Step 1: Define the desired precision level.
Step 2: Estimate the variance.
Step 3: Estimate the cost of estimation or monitoring.
Step 4: Define the permissible error.
Step 5: Define the confidence interval.
Step 6: Determine the number of strata.
Step 7: Estimate the number of plots.

Step 1: *Define the desired precision level* Typically, to estimate the number of plots needed for measuring and monitoring at a given confidence level, it is necessary to first estimate the variance of the variable (e.g. carbon stock of the main pools, trees in an afforestation or reforestation project or soil in a cropland management project) in each stratum (IPCC 2003). This can be accomplished either by using existing data from a project similar to the one yet to be implemented (e.g. a forest or soil inventory in an area representative of the proposed project) or by conducting pilot study from an area representative of the proposed project.

Carbon inventory requires reliable estimates, which means the values are precise and accurate. Higher the level of precision, larger the sample size and higher the cost. The level of precision should be determined at the beginning of the project, which could vary from ±5% to ±20% of the population mean. However, a precision level of ±10% of the true value of mean at a confidence interval of 95% is normally adequate, although a range of ±5% or even ±20% is also often employed.

Step 2: *Estimate the variance* An estimate of variance of the carbon stocks is required for each stratum, which could be obtained from studies conducted in a region with conditions similar to those for each proposed project activity. If such estimates are not available, pilot studies may be required in locations close to the project area. Such a study involves the following steps:

- Identify an area near the project area with conditions similar to those for the proposed project activities (e.g. tree plantation, agroforestry or protected forest).
- Conduct field studies by selecting a few small sample plots in the selected vegetation type. Measure the relevant tree or non-tree parameters such as DBH, tree height and weight of shrub or herb biomass (as described later in the chapter as well as in the following chapters).

- Calculate the mean and variance from the data collected from the pilot study.

Step 3: *Obtain cost estimates for monitoring* While conducting the preliminary field studies, keep notes of all the costs involved in traveling, laying plots, making measurements and calculations and any other expenses. Using these values, estimate the cost of sampling a plot in a given stratum. If cost figures are available from other similar studies, use them.

Step 4: *Permissible error* Estimate the permissible error in the mean carbon stock value estimates. Usually, the permissible error value can be taken as ±10% of the expected mean carbon stock.

Step 5: *Confidence interval* Choose a value of 95% confidence level.

Step 6: *Number of strata* Select the number of strata for the project activity (refer to Section 10.3).

Step 7: *Estimate number of plots* Calculate the required number of plots using the following formulae:

$$n = \left(\frac{t_\alpha/2}{A}\right)^2 \left(\sum_{i=1}^{N_i} W_i s_i \sqrt{C_i}\right)\left(\sum_{i=1}^{N_i} W_i s_i \sqrt{C_i}\right)$$

where n sample size (number of sample plots required for monitoring), t_α value of Student's t statistic for $\alpha = 0.05$ (implying a 95% confidence level), N_s total number of strata designed, N_i number of potential sample units (permanent sample plots in the stratum level), S_i standard deviation in stratum i, A permissible error in the mean, C_i cost of selecting a sample plot in stratum I, $W_i = N_i/N_s$.
The number of plots shall be allocated among the strata:

$$n_i = n \cdot p_i \ and \ p_i = \left(W_i s_i \sqrt{C_i}\right) / \left(\sum_{i=1}^{N_i} \cdot W_i \ s_i \sqrt{C_i}\right)$$

where n_i is the number of samples to be allocated in stratum i.

The allowable error is a value on per-plot basis estimated as ±10% of the mean biomass carbon stock per plot. The sample size determined during the project development phase could be modified during the project monitoring phase if new information necessitating a change in sample size becomes available.

Sampling error The term sampling error is used to express the difference between a population mean and the mean of a sample drawn from that population. This error is caused by sampling. If no error exists, the sample mean will be exactly equal to the population mean. When plot-level measurements are scaled up, sampling errors occur because conditions across the larger area vary whereas measurements are made only at the sample locations. The average conditions within the selected sample plots seldom coincide exactly with the average conditions within the entire area of interest. Sampling errors (using random sampling designs and unbiased estimators) are only random and can be reduced by increasing sample size. The relation between sampling error, population variance and sample size is generally known: increasing sample size leads to higher precision, and heterogeneous populations (forests with diverse vegetation) require a larger sample size to obtain the desired precision.

10.7.4 Size of the Plot

A carbon inventory requires that both size and number of sample plots be decided. Plot size too has implications for the cost of carbon inventory or monitoring. Larger the plots, lower the variability between two samples. Therefore, plot size depends on the extent of variation among plots and the cost of measurement. According to Freese (1962), the relationship between plot size and its coefficient of variation (CV) is given by the following equation:

$$\mathrm{CV}^2 = \mathrm{CV}_1^{\,2}\sqrt{(P_1/P_2)}$$

where P_1 and P_2 represent plot sizes and their corresponding coefficients of variation.

Increase in the plot size reduces variation among the plots and leads to smaller number of plots. Very often, the number of plots is selected based on expert judgment about the size of trees, size of project area and variations in stand density. Some illustrative numbers for different tree sizes and types of vegetation are given below. Diverse vegetation (natural vegetation) requires larger plots and homogenous vegetation (plantations of uniform age and density) requires smaller plots. In Table 10.2 Pearson et al. (2005b) suggest plot sizes for trees of different size.

Error in sampling that affects precision could occur because of variation in sample units and a large project area, measurement techniques or instruments, models or allometric equations and other errors. The relation between sampling error, population variance, and sample size is as follows (IPCC 2003):

- Increasing sample size increases precision
- Heterogeneous populations (i.e. those with large within-population variation) require larger samples to reach a given level of precision
- Where area proportions are to be estimated, sampling errors depend not only on sample size, but also on the proportion itself.

Table 10.2 Suggestion of plot size

Stem diameter (DBH, cm)	Circular plot (radius, m)	Square plot (m)
<5	1	2×2
5–20	4	7×7
21–50	14	25×25
>50	20	35×35

10.8 Preparation for Fieldwork and Recording of Information

Estimation and monitoring of biomass in land-use systems involves measurement of plant-based parameters such as DBH and height of trees and weight of non-tree biomass. It is important to plan fieldwork well in advance for efficient use of staff and time, and necessary to procure all the background information before launching field studies. Field studies require:

- Trained staff
- Background information
- Instruments and materials for measurement
- Arrangement for collection of plant samples
- Formats for recording data

Trained staff Field studies require at least one trained person and one or two field assistants. The trained person takes the measurements and records them in the formats provided; the field assistants help in laying plots, holding the measuring device (tape, pole or scale), establishing the boundary and putting in corner pegs. It is always desirable that whosoever records the data in the field should also be the one to enter the data into the computerized database.

Background information Before embarking on field studies, it is important to obtain all the relevant background information, which will help in laying plots or taking measurements. Such background information could be obtained from the project office, land survey or forest departments, local government offices and local communities. It is particularly important to collect all the available maps and prepare a map showing the project area and boundary, baseline land-use systems and characteristics and the project scenario activities. The type of background information required includes:

- Projection of location maps showing latitude and longitude, topographic sheets, forest map and soil maps
- Names for land-use systems and their location and area
- Elevation, topography, broad soil type and rainfall
- Proximity to human settlements, roads, urban centres, markets
- Land tenure or ownership
- Livestock population and grazing locations
- Past land-use changes and features
- Data on afforestation, reforestation, soil and water conservation, i.e. programmes or activities implemented or proposed
- Sources of fuelwood and timber
- Socio-economic and demographic features

Instruments and materials for measurement The materials and instruments required for field studies on above-ground biomass estimation are given in the box below. The material required could most often be procured locally. The materials used should be of durable quality and the scales and measuring instruments validated or calibrated.

Arrangement for collection of plant samples Cloth and polythene bags are required for collecting plant samples for obtaining fresh as well as dry weights. A balance is required for taking fresh weights in the field. The dry weight is obtained by drying a sample of known weight in an oven to constant weight.

Data recording formats Formats vary for different plant forms such as trees, shrubs and herbs and need to be standardized for both field recording and for entering into the database. Sample formats are provided in Section 10.11.

- *Long measuring tape* (5, 30, 50 m)
- *Fine measurement tape* for measuring DBH (1 or 1.5 m)
- *Rope and pegs* for marking boundary and corner points
- *Paint and brush* for marking the point at which DBH is measured
- *Aluminium tags* for marking trees
- *Global Positioning System* (GPS)
- *Clinometers* for measuring tree height
- *Slide calipers* for measuring DBH of small stems
- *Balance* for weighing shrub, woody litter and herb layer biomass
- *Cloth bags* for sampling harvested or litter biomass for dry weight estimation
- *Metallic frame* for sampling for herb layer biomass (1 × 1 m)
- Data recording sheets and pencils

10.9 Sampling Design

Sampling design aims at locating the sample plots in each of the selected stratum. The soil, topography, water availability and status of vegetation vary spatially within a land-use category, area proposed for the project activities or even in the area brought under a given project activity. Trees, biomass stock and growth rate are not distributed uniformly in a given project area or even for a given project activity, and the location of sampling plots could determine the biomass stock or growth rate estimates. The project staff may be biased in locating sample plots in spots with good tree or grass growth to obtain higher biomass stock values. Sampling techniques ensure unbiased selection of sites for laying out sample plots in the field. The main purpose of adopting a sampling design is to avoid bias in locating the sampling plots in land-use systems in both baseline scenario and project scenario. Different sampling designs for laying out sample plots for vegetation studies are as follows.

Selective sampling Selective, subjective or purposive sampling design is used for vegetation studies to assess the biomass carbon stock or roundwood production at some selected locations or as part of some projects. Although the time and effort required for locating and laying out sample plots selectively are minimal (Kangas and Maltamo 2006), the biomass estimate may not be reliable, reflecting a ground value that is not representative of the site. The error is likely to be high. Purposive sampling could be adopted, for example, to estimate the above-ground biomass of a project activity by laying plots close to and away from a village settlement to assess the impact of grazing or fuelwood extraction.

Simple random sampling To apply the simple random sampling technique, convert the project area into a large number of equal-sized grids. In this method, the sample plots are laid out randomly to avoid bias in locating the plots (Fig. 10.4). Random sampling ensures that each point or grid in the inventory area has an equal chance of being included in the sample. Further, the position of one plot has no influence on the position of other plots. Randomization makes it possible to obtain unbiased estimates of variability as well as the mean per unit area. However, randomized sampling layouts are not very convenient for field staff in locating the plots during periodical monitoring (Myers and Shelton 1980). Simple random sampling method is not often adopted out of consideration to the heterogeneity of the population or the project area because it is based on the premise that the population is homogenous, however, the method could be adopted when no prior information is available about the project area. All the area under project activity is therefore considered as one unit and the heterogeneity of soil, topography or other features is not considered.

Stratified random sampling The features and benefits of stratification were described in Section 10.3. Stratification leads to efficient sampling and reduction of standard error. Each stratum can be considered as a subpopulation. In this technique, the project or activity area is stratified based on key features such as soil quality, topography, level of degradation and vegetation status and particularly the density and size of the trees. Area under each stratum is subdivided into a large number of equal-sized grids and the grids are numbered. The sample plots are chosen randomly among the grid numbers of each stratum, using the approach adopted for simple random sampling. The steps involved in stratified random sampling are as follows:

- Stratified random sampling becomes more effective with increasing homogeneity within each stratum.
- The approach is to implement the sampling procedure separately in each of the strata and then pool the information for a given project activity or land-use category.
- Stratified sampling avoids the possibility of large differences between strata contributing to the sampling error. The stratification of the sample leaves only the relatively small variation within each stratum to be reflected in the sampling error.

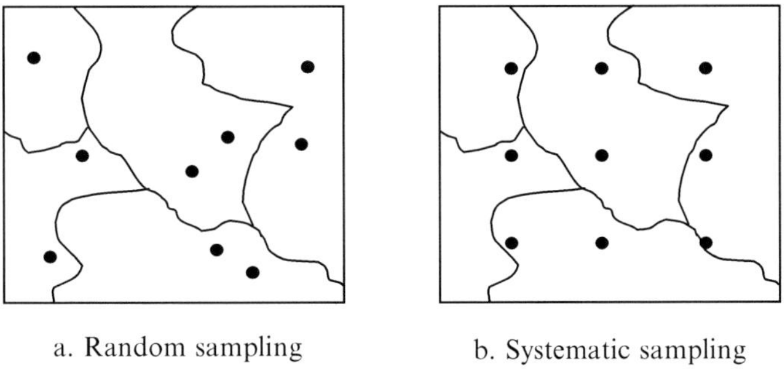

Fig. 10.4 Layout for simple random sampling (*left*) and for systematic sampling (*right*)

Systematic sampling In systematic random sampling, the sample plots are placed at fixed intervals throughout the project area for a given activity. As the term implies, sample plots are not randomly distributed over the inventory area but arranged in a systematic pattern (Fig. 10.4). An important feature of systematic sampling and layout is that the position of the first plot, which is chosen at random, determines the positions of all the subsequent plots. According to Myers and Shelton (1980), the main advantage is that this approach is simple and can be adopted even in the absence of the maps. The regular spacing and systematic layout does tend to give convenient patterns for travel and fieldwork.

The disadvantages include: (i) the regular spacing of sample units might coincide with some cyclic fluctuation in the vegetation being sampled; (ii) dependence on the location of the first sampling unit; and (iii) the difficulty in estimating the variability of the population from systematic samples.

10.10 Location and Laying of Sample Plots

This section presents an approach to locating and laying out sample plots in the field in different land-use systems. The criteria in locating the plots are as follows:

- Plots located must be representative of the land-use system.
- Plots must be located in an unbiased way within the land-use system, except those for estimating leakage (Chapter 6).
- Plots should be accessible to investigators for measurement and monitoring.

The selected number of plots needs to be located and laid in the carbon inventory area in an unbiased manner, given the variations in soil, topography, vegetation, etc. Sample plots are required to be located and laid out during the project-development phase as well as project-monitoring phase. The main approach involves: (i) fixing the number of plots for each stratum or project activity; (ii) selection of sampling design; and (iii) location of the sample plots in the carbon inventory area converted to grids in each sampling stratum. The sample plots can be laid without any bias as follows:

Step 1: Select and stratify the project area or the area under each activity.

Step 2: Obtain a map of the total project area and convert it into grids of appropriate size depending upon the area under a project activity or baseline land-use system. The grids could be 10 × 10m to 100 × 100m. It is desirable to make the grid size larger than the size of the sample plots. Further, the number of grids is usually several times the number of sample plots.

Step 3: Numbers the grids 1 to n, where n is the total number of grids.

Step 4: Select the number of sample plots for each land-use system under the baseline scenario and project activity stratum by using the methods described in Section 10.7.3.

Step 5: Select the sampling design: simple random sampling, stratified random sampling or systematic sampling.

Step 6: Locate the sample plots in the carbon inventory area using the sampling design adopted (using the steps described in the following section).

(i) *Simple random sampling*

- Randomly pick as many grid numbers as the number of sample plots, using a table of random numbers or by drawing lots. For instance, if five tree plots are to be selected, pick five random numbers.
- Ensure that the randomly drawn plots do not all fall into a single cluster, which is rare.
- Locate tree plots in the grids selected in the field with respect to some permanent visible landmark and mark the boundary of each tree plot or use GPS.
- Prepare and store a map with all the details, including the location of sample plots marked on it. If GIS is available, it can be very useful.

(ii) *Stratified random sampling*

- Stratify the land-use system or project activities into homogeneous units.
- Select the stratum.
- Adopt for each stratum the procedure given for simple random sampling.
- Repeat the procedure for laying plots for the next stratum and continue until all the strata are covered.

(iii) *Systematic sampling*

- Stratify the land-use system into a number of homogenous units.
- Obtain a map showing the grids in each sampling stratum and estimate the total number of grids for each stratum (N): e.g. 200 grids with a total project area of 40 ha (which is worked out below as an illustration). The plot numbers and the locations of the sample plots are marked on the grid map of the carbon inventory area.
- Calculate the sampling interval "k" by using the following equation:

$$k = N/n$$

where k = sampling interval of grids or plots = 200/5 = 40, N = total number of grids representing a given strata (200) and n = number of sample plots (quadrats) to be selected (e.g. 5).

- Draw a random number smaller than k (smaller than 40 in this example), say 25.
- Select and mark the first grid based on the random number.
- The first sampling grid number is 25.
- The second sampling grid or plot = sampling interval k (40) + first sampling grid (25) = 65.
- The third sampling grid or plot = sampling interval k (40) + second sampling grid (65) = 105.
- Repeat the procedure for the remaining number of sample plots.

Marking the plot in the field The plot numbers and the locations of the sample plots are marked on the grid map of the carbon inventory area. These grid numbers have to be located in the field for long-term periodical monitoring of vegetation. The following steps could be considered to facilitate the process:

Step 1: Use the carbon inventory project area maps with sample plots marked on the grid map along with their geographic coordinates (latitude and longitude).

Step 2: Locate the sample grids on the ground using GPS points from the map or using any permanent visible landmark in the field.

Step 3: Mark the corners of the sample quadrats on the ground using pegs or any other permanent marking arrangements for long-term periodical monitoring. To avoid any special treatment to the permanent sample plots, it may be desirable to hide the corner points of the quadrats.

Step 4: Use GPS positions of the quadrat corners for long-term periodic visits to avoid any bias in treatment of the vegetation in sample plots. The boundary line of the plot should be marked with a rope or coloured chalk powder during measurement.

Marking of tree, shrub and herb quadrats Tree sample plots or quadrats are normally several times larger than the shrub quadrats, which are several times larger than the herb or grass quadrats:

- Measure and mark the corners and boundary of the tree quadrats in the field.
- Mark the shrub quadrats within each of the tree quadrats, normally at two opposite corners, keeping two shrub plots per tree quadrat.
- Mark the herb or grass quadrats within the shrub quadrats at the opposite corners, keeping two herb plots per shrub plot.

Location and layout of the tree, shrub and herb quadrats could be along the following lines (Fig. 10.5).

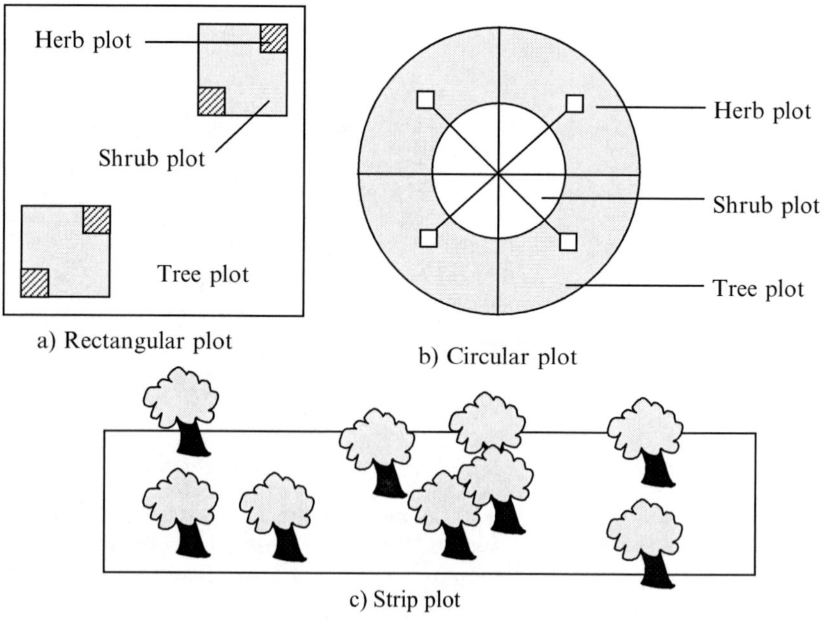

Fig. 10.5 Shape or type of sampling plots: (a) rectangle, (b) circle and (c) strip

Table 10.3 Size and number of plots for different land-use systems or project activities under baseline and project scenario. (From Pearson et al. 2005b.)

	Trees		Shrub		Herb/grass		Soil	
Land use system	Size of plot (m)	No. of plots	Size of plot (m)	No. of plots	Size of plot (m)	No. of plots	Size of plot (m)	No. of plots
Natural forest or heterogeneous vegetation	50×40 50×50	5 4	5×5 5×5	10 10	1×1 1×1	20 20	1×1 1×1	20 20
Plantations with homogenous vegetation or uniform species distribution and density	50×20 or 40×25	5	5×5	8	1×1	16	1×1	16
Savannah or grassland or rangeland with few trees	50×40	5	5×5	10	1×1	20	1×1	20
Degraded forest or barren or fallow land	50×40	5	5×5	10	1×1	20	1×1	20

An illustrative sample size and number of tree, shrub and herb plots for different land-use systems are given in Table 10.3. Such sample sizes are often adopted.

10.11 Field Measurement of Indicator Parameters

Estimation of carbon stock in biomass or its growth rate requires measurement of indicator parameters such as tree height and GBH. These parameters are measured in the field through a sampling design. Field measurements for biomass carbon assessment are required at:

- *Project development phase* for baseline scenario land-use systems and proposed project scenario activities
- *Project monitoring phase* for baseline scenario land-use systems and implemented project scenario activities

Above-ground biomass is estimated in any typical land-based projects for trees, shrubs and herbs/grasses. The biomass of these plant forms is measured using the following steps:

Step 1: Decide on the sample size; locate and mark the sample plots for trees, shrubs and herbs on the ground (Section 10.7 and 10.11).

Step 2: Select the parameters for tree, shrub and herb biomass (Section 10.6) and procure all the material required for field studies.

Step 3: Measure the parameters for trees, namely species, height, DBH and status or features.
Step 4: Measure the parameters for shrubs, namely height, DBH and weight of the woody and non-woody biomass.
Step 5: Measure parameters for herbs/grass, namely species, number of plants, weight of the plants in the sample plots.
Step 6: Record all the parameters in the standard format for trees, shrubs and herbs/grass.

Steps 1 and 2 have already been described in earlier sections. Steps 3–6 are described in Sections 10.11.1–10.11.5. These steps largely focus on measurement of different parameters as indicators of plant biomass.

10.11.1 Above-Ground Biomass of Trees

Trees are woody perennial plants having a single, usually elongated main stem with few or no branches on their lower part. Trees could be large or mature (DBH > 30 cm), medium-sized or growing (DBH 10–30 cm) or regenerating seedlings (DBH < 10 cm). A plant belonging to a tree species is considered for measurement in tree quadrats if it is taller than 1.5 m and its DBH is greater than 5 cm (a girth of about 15 cm). The height and DBH class to be considered in the tree quadrat will vary with the project type and age of the stand. In the case of old, mature forests with large trees, a tree could be defined to have a DBH of greater than 30 cm. Locate the tree plots in the field and measure the DBH, height and other parameters for all the trees using the procedure described in this section.

Parameters to be measured include species, number of stems, DBH, height, status of regeneration, state (living, dead and standing, dead and fallen) and extent of damage to crown.

Frequency of measurement for trees will vary from 1 year for fast-growing tree species to 5 years for slow-growing naturally regenerating trees (refer to Chapter 4 for details).

10.11.1.1 Measuring

DBH DBH is easy to measure and verify. It requires only a measuring tape and a marker. DBH is measured using the following procedure:

- Mark a point 130 cm above the ground on the tree trunk.
- Place the measuring tape around the trunk at the 130 cm mark.
- Measure and record the DBH or GBH in centimetres:
  - If a tree has multiple shoots, count and measure GBH/DBH for all shoots.
  - If the tree is large, girth is normally measured with a measuring tape.

- If the tree is young and slender, measure the DBH with a slide caliper.
- If the tree is on the boundary, include it for measurement in the sample plot only if more than 50% of its girth is inside the plot.

A tree could have multiple shoots and/or crooked trunks, could be growing at an angle, and could be on a sloping hill. The measurement technique for irregularly shaped trees and different land conditions is illustrated in Fig. 10.6.

10.11.1.2 Height

Tree height normally refers to total tree height defined as the vertical distance from the ground level to the uppermost point. Tree height is also often referred to as merchantable height since many allometric equations are derived for this height. Height is measured for all the tree stems for which DBH is measured (Commonwealth of Australia 2001). Unlike DBH, measurement of tree height is difficult for tall

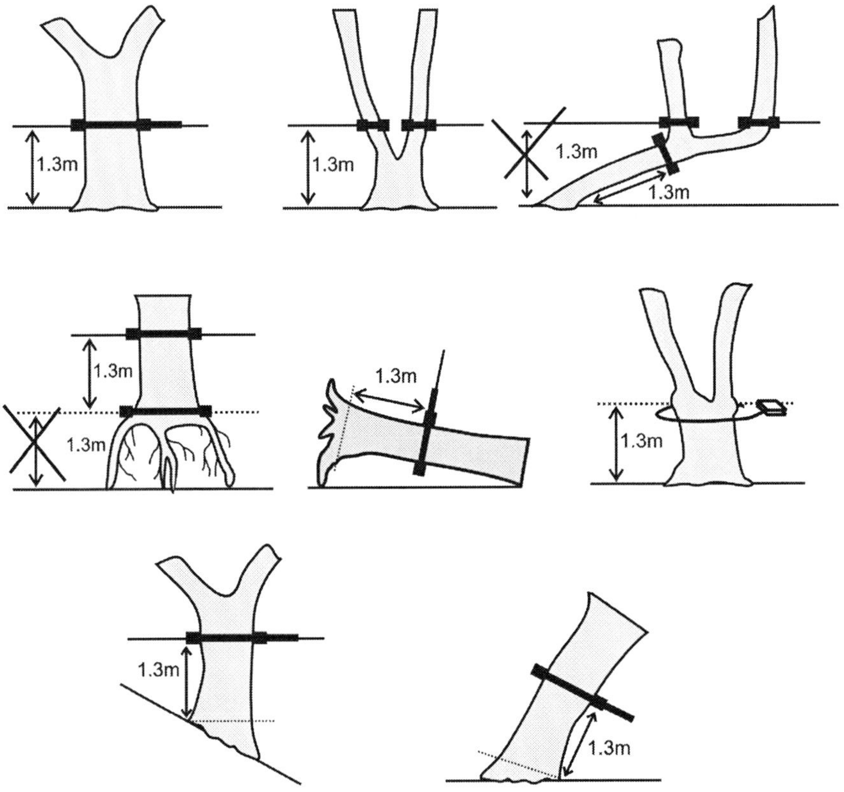

Fig. 10.6 Measuring DBH or GBH for trees of different shapes and forms

trees, especially in a dense forest or plantation with tree close together and overlapping crowns. Height can be measured using different methods.

(i) *Measurement using instruments* Tree height can be measured using various instruments or even using a measuring tape. However, measuring the height of individual trees with overlapping tree crowns and trees in a dense forest or plantation poses a challenge for measurement even using instruments. Trees taller than 5 m can be measured using a graduated height-stick by holding the stick against the side of the tree. Clinometer is one of the instruments used for measuring the height of trees but it is not suitable for dense vegetation where visibility is limited. Mark out a spot 10 m from the tree from which the tree can be viewed using a clinometer. If necessary, move beyond 10 m. If the plot is located on a steep slope, view the tree from across the slope to maintain the required distance. Sighting the tree through the clinometer, align the centre line with the base of the tree (ground level on the upside slope) and record the reading on the percent scale (base angle %). Next, aim the clinometer at the top of the tree and record the reading on a percentage scale. Calculate the height using the following equation:

$$\text{Height (m)} = \frac{[\text{top angle (\%)} - \text{base angle (\%)}] \times \text{horizontal distance}}{100}$$

(ii) *Height classes* Trees can be grouped into different height classes (e.g. 0–5, 6–10, 11–15, 16–20 m). These classes can be used as reference height classes to get an approximate estimate of the height of trees in a plot. Trees are observed during fieldwork and categorized into these height classes. Field investigators with a little practice and experience can estimate the height class of a tree by mere observation and place it in the appropriate class.

(iii) *Height equation* Height for a given tree species is correlated with its DBH. A regression equation can be developed for a given species by measuring the height and DBH of atleast 30 sample trees of different heights. Using the height estimation equation, height can be estimated from DBH data for a given tree species. The equation could be along the following lines.

$$\text{Height (H)} = a + bD$$

where D = DBH, a = constant, b = regression coefficient.

For example, Chaturvedi (2007), based on the measurements of more than 4,000 sample trees of teak (*Tectona grandis*) of different age classes in Western Ghats, India, estimated the relationship between DBH and height, by considering height as the dependent variable and DBH and age as independent variables.

$$\text{Height} = 35.02{*}(\text{DBH})^{0.66} \qquad R^2 = 0.73, N = 4002$$

$$\text{Height} = 2.933 + 34.093{*}\text{DBH} + 0.05{*}\text{Age} \qquad R^2 = 0.71, N = 4002$$

Height data are required for all the tree species during the project development phase as well as project monitoring phase. The following approach could be adopted for measuring the height of trees during these two phases.

Project development phase During the project development phase, height of trees in the baseline scenario could be measured using a measuring tape or clinometer, since only a few trees may have to be measured. Alternatively, tree height class method could be adopted for forests or plantations with dense tree population.

Project monitoring phase Multiple methods could be adopted for measuring tree heights during the monitoring phase:

- During the seedling or younger stage when the trees are shorter than, say, 5 m, all trees in the sample plots could be measured using a simple graduated pole or measuring tape.
- When the trees grow taller, either height–DBH equations could be developed by measuring trees of different heights and DBH or height-measuring instruments could be used.

Regeneration status It is important to know whether a tree has regenerated or grown from a seed or sapling planted deliberately or has grown naturally as a part of the project activity. Such a distinction – trained field staff can make out the difference – enables the contribution of each to the carbon stock to be estimated separately. This information should be recorded in the data format for each of the tree species.

Status of trees The following information regarding the health of trees should be recorded, which will be useful in estimating the quantities of biomass from living trees and from deadwood:

- If the crown is damaged, the percentage of damage
- Whether the tree is dead and standing or dead and fallen

Tagging of trees Perennial trees may have to be measured periodically over a number years or even decades in forestry projects. Therefore, it may be desirable to mark or tag them to make it easier to find them and to ascertain their species and number. This purpose could be achieved by fixing aluminium or other metallic tags to the trees.

A suggested format for recording the data is shown in Table 10.4.

10.11.2 Shrubs

Shrubs are woody plants that are usually short, often less than five metres, with several stems arising from the base and lacking a single trunk. Shrub plots include

Table 10.4 Suggested format

Location: GPS reading		Land-use system: stratum		Plot no: size of the plot	Investigators: date	
S. No 1	Species name	Tree number	GBH of stem (cm) 1 2 3 4 5	Planted or regenerated	Height (m)	Status of crown[a]

[a]Indicate the percentage crown cover present or damaged

shrub species as well as younger trees with DBH lower than what is defined for the trees in the tree plots. Shrub plots are located inside the tree plots (Fig. 10.5).

Parameters to be measured include species, number of stems, DBH, height and weight of the shrub biomass from the sample plot.

Frequency of measurement for shrub vegetation could vary based on vegetation types. Normally, adopt the same frequency as that used for the tree plot measurements.

Demarcation of the shrub plots and the boundary can be undertaken by following the method of laying out shrub plots described in Section 10.10. Shrub plots are usually located at two opposite corners of a tree plot (see cover of the book). If a shrub is on the boundary of the plot, treat it as part of the plot so long as more than 50% of the shrub crown is inside the plot.

Procedure for measuring shrubs and young trees in shrub plots The following steps could be adopted to measure the parameters in the shrub plots:

Step 1: Locate and number the shrub plots in each of the tree plots.
Step 2: Start from one corner of the shrub plots and record the indicator parameters and mark the plants after measurement with a chalk or paint.
Step 3: Record the species and the number of shrub plants under each species.
Step 4: Measure the height of the tree using the methods described for trees.
Step 5: Measure DBH of all trees taller than 1.5 metre in the shrub plot; if multiple shoots are present, record DBH for all the shoots.
Step 6: Record the name, height, DBH and other features for each shrub plant in the format provided.

Procedure for measuring non-tree vegetation Vegetation other than trees could include annual or perennial shrubs as well as very young seedlings (shorter than 1.5 m). Estimation of non-tree biomass in shrub plots follows the same method as that for annual and perennial shrubs. Tree seedling should be excluded from harvest procedure.

Biomass of annual shrubs is estimated by cutting the shrubs in the shrub plot, one species at a time, and recording the fresh weight of all the plants. Dry weight of the biomass can be estimated by taking a known quantity (0.5–1.0 kg) of a plant sample and drying it to constant weight in the oven.

Biomass of perennial shrubs is estimated by harvesting the perennial shrub plants in the shrub plot, again species by species, and estimating the fresh and dry weight of the plants. However, if the shrub species is yielding any economically valuable product, such shrub species need not be harvested or a few representative shrub plants could be harvested to get the mean weight for the species. The mean weight of sample shrubs harvested can be extrapolated to the whole plot.

Periodic monitoring of shrub and tree biomass Periodic monitoring of shrub biomass could be through harvesting, using the "permanent plot". However, for harvesting, each time select a plot adjacent to the plot harvested before so that measurements are comparable and the impact of earlier harvest is avoided. A suggested format for recording the data is shown below.

Location: GPS reading	Land-use system: stratum	Tree plot no: shrub plot no: size of the plot		Investigators: date
S. No. Species	Diameter (cm)		Height (m)	Biomass – Fresh weight (kg)
1	DBH1 DBH2 DBH3			

10.11.3 Herbs

Herbs are non-woody plants that usually die at the end of the season. Herb-layer biomass includes all annual plants, regenerated saplings and grass biomass. Herb-layer plots are usually small (1×1 m) but more numerous. Biomass in the herb layer is part of the annual carbon cycle and is estimated by harvesting during the peak growth period. Methods for estimating biomass from grassland are given in Section 10.11.4.

Parameters Species name, number of plants and fresh weight of standing herb biomass are the parameters to be recorded.

Frequency of estimation The biomass of herbs is recorded annually during the peak growth period.

Demarcation of herb plots and the boundary Herb plots are usually 1×1 m and marked at the two opposing corners of each shrub plot (Fig. 10.5).

Measurement of herb vegetation Measuring herb vegetation involves recording the species name and harvesting the herb biomass to determine its weight. The following steps could be adopted:

Step 1: Record the species name and number in each herb plot. The percentage of ground covered by herbs in the plot could also be recorded, by species, based on visual observation.

Step 2: Select the peak month of biomass growth for harvesting. Alternatively, adopt the method described for grasses in Section 10.11.4. Cut the herb plants, by species, in each herb plot.

Step 3: Take the fresh weight of the herb biomass, again by species.

Step 4: Estimate the dry weight by taking a small sample of fresh herb biomass and drying it to constant weight in an oven.

If there is any ban on harvesting certain herb species, avoid harvesting those species. Also avoid harvesting saplings or seedlings of valuable tree species.

10.11.4 Grass Production

Grasslands are characterized by dominance of grass species, with a few trees or total absence of trees or other perennials. Estimates of above-ground grass production, which is part of the annual carbon cycle, may not be very relevant to carbon inventory projects or greenhouse gas inventory programmes. However, estimating grass production is essential for grassland reclamation projects. The method for estimating grass production is similar to that for annual herbs, and the following steps could be adopted:

Step 1: Select the grassland category or grassland improvement project activity and stratify the land area.

Step 2: Adopt standard sampling procedures: decide on the size and number of plots and locate them in the field using the procedure described for herb plots.

Step 3: Select four to five tree quadrats the same way tree or shrub quadrats are selected, or select the tree quadrat itself in forests or plantations

Step 4: Divide each shrub or tree quadrat into 12 subplots of 1 × 1 m to represent 12 months of a year and mark and fence the plots.

Step 5: Harvest the grass by clipping the above-ground biomass in subplot 1 of quadrat 1 and determine the fresh and dry weight of the grass biomass harvested.

Step 6: Repeat the harvest procedure for subplot 1 of all the quadrats the same month, estimate the fresh and dry weight of grass biomass (grams/square metres) and convert it to per hectare estimate of grass production (grams/hectare) in terms of dry weight.

Step 7: Next month, repeat the harvest and biomass estimation in subplot 2 of each of the four or five quadrats, estimate fresh and dry weight, and estimate the grass production (dry weight) for that month in terms of grams/hectare.

Step 8: Repeat the process monthly to cover the remaining months. Compile the monthly production of grass biomass (grams/hectare), calculate the average biomass for all the grass-growing months, and select the month in which grass production is maximum to assess the productivity of that grassland category. Harvesting could be avoided when no biomass growth is likely.

10.11.5 Measurement of Palms and Lianas

Palms Palms are a group of large tree-like plants with a single, tall, unbranched stem with a crown of large fan-shaped leaves. The height alone is adequate for estimating the biomass of palms since, in palms, biomass is more closely related to height than to DBH and the biomass equations are based only on height. If palms are present in the sample plots, adopt the following steps (Pearson et al. 2005b):

- Measure the height of the palm from base up to the point where the stem is no longer visible.
- If the plot has to be re-measured, mark it as appropriate.
- Record the plot number, palm number and height.
- If possible, harvest 20–30 palms of different heights to get the mean weight of the palm or classify them by height.
- Use the biomass estimation equations available for palms if required.

Lianas Lianas are woody perennial climbing plants with very long stem, which grows around trees right up to the top where there is more sunlight. Lianas are difficult to measure since they are often long, winding and cross the plot boundary at many places (Pearson et al. 2005b). Include lianas only if they are likely to constitute a significant proportion of the biomass of the plots. It is difficult to estimate the biomass of the lianas without harvesting them, and no biomass estimation equation is available for lianas.

10.12 Recording Data and Compilation

Data recording formats have been developed for tree, shrub and herb species in sample plots. These formats are largely for use in the field. The data entered in these formats in the field need to be verified and entered into a database for analysis. Some of the precautions and steps to be followed to ensure correct recording in the field and proper compilation for obtaining reliable estimates of biomass are as follows:

- Use the appropriate data entry format for trees, shrubs and herbs.
- Remember to enter the location name, date, plot number, vegetation type and name of the field investigator.
- Enter and verify GPS readings of the plots.
- Enter and verify the units of height, DBH and weight.
- Ensure that all the relevant data-recording cells in the formats are filled before leaving the field.
- Verify the data recording formats as soon as possible after returning from the field for any corrections or conversion of traditional units of measurement to standard units such as the SI units.
- Codify any entry as required by converting qualitative information into codes, for example, presence or absence (0 or 1), land-use systems (1: agriculture, 2: grassland, 3: settlement, 4: tree plantations).
- Develop a user-friendly data entry system for computer analysis and for archiving data.
- Verify all the data entered and store them in a database.

The procedure for analysing the data to estimate above-ground biomass is presented in Chapter 17 and methods to estimate uncertainty in Chapter 18.

10.13 Long-Term Monitoring for Above-Ground Biomass

Long-term observations and monitoring are central to the study of almost every important ecological concept and every environmental issue. Long-term monitoring is extensively adopted in ecological studies to understand ecosystem changes, vegetation succession, carbon dynamics, biodiversity changes and other ecological processes (Franklin 1989). Long-term monitoring is critical to carbon inventory since carbon gains and losses occur over long periods, spanning decades and centuries. Above-ground biomass accumulates over decades and centuries, although it may peak sometime during that period. The peak time varies with forest type and plantation species. However, carbon stock could be lost from different land-use systems within a short time because of disturbance, especially in the form of fire, land conversion and harvesting. Long-term monitoring in the context of carbon inventory for above-ground biomass is required for such projects or situations as:

- Land conversion from forests or plantations to degraded land, cropland or grazing land
- Afforestation and reforestation of degraded lands to sequester carbon
- Avoided deforestation to conserve forest carbon sink
- Roundwood production and bioenergy plantation programmes
- Agroforestry and shelterbelt programmes
- Land reclamation projects

A long-term monitoring plan should be developed and incorporated into the project during the planning and project development phase and adopted during the post-implementation phase. This section briefly presents the methods for and steps in long-term monitoring of changes in above-ground biomass.

Methods for long-term monitoring of above-ground biomass Many of the methods mentioned in Chapter 9 could be adopted for long-term monitoring. The two promising methods are permanent plot method and remote sensing techniques. Because remote sensing techniques with practical applications in land-based carbon mitigation projects, roundwood production programmes and national greenhouse gas inventory are still evolving, the permanent plot method is the most suitable one for long-term monitoring on account of the same merits mentioned in Chapter 9, namely cost-effectiveness, suitability to small and large projects and minimal staff and training requirements. The steps presented for plot method earlier in this chapter and described in Sections 10.1–10.13 are applicable to long-term monitoring. Here, only a few special features to be considered while planning and implementation of long-term monitoring studies are presented.

(i) *Sampling* Sampling methods involving selection of size, number and shape of the plots and the design for locating the plots are the same as those described earlier in Sections 10.6–10.10. For long-term monitoring, the quadrat method with a large plot size (e.g. 50 × 50 m) is desirable for revisits and measurements.
(ii) *Location and layout of plots* The plots should be located using the chosen sampling design using (Sections 10.9 and 10.10). For long-term monitoring, it is important to mark the plots in the field as well as on the map using GPS readings and with reference to any permanent landmark for easy identification on the ground.
(iii) *Recording and archiving of data* It is very important to develop formats for recording data in the field as well as entering them in a database.
(iv) *Staff and training* The staff involved in field and laboratory studies as well as data entry and analysis require training. In the long term, staff turnover in any project is quite likely. Therefore, it is important to maintain detailed guidelines and manuals on field and laboratory studies as well as data entry and analysis protocols.

10.14 Conclusions

Above-ground biomass is an important carbon pool for (i) all land-use categories for national greenhouse inventory, (ii) carbon mitigation projects, particularly tree-based projects, and (iii) roundwood production programmes. Among the multiple methods for estimating above-ground biomass, the plot method is covered in detail because it is simple, reliable, widely applicable and cost-effective. The critical components of the procedure are sampling and field measurement. The data gathered using the plot method will enable estimation of above-ground biomass stock, growth rate and stock changes. Above-ground biomass can be estimated for trees, forests, plantations, grasslands and croplands using the plot method. The procedure for analysis and estimation of carbon stocks in above-ground biomass is presented in Chapter 17. All the methods and steps presented in this chapter for carbon mitigation and roundwood production projects are fully applicable to forest inventories as well as carbon inventory for national greenhouse gas inventories from land-use categories. Adoption of permanent plot methods will enable long-term and periodic measurement and estimation of carbon stocks over any selected period.

Chapter 11
Methods for Below-Ground Biomass

Below-ground biomass is defined as the entire biomass of all live roots, although fine roots less than 2 mm in diameter are often excluded because these cannot easily be distinguished empirically from soil organic matter. Below-ground biomass is an important carbon pool for many vegetation types and land-use systems and accounts for about 20% (Santantonio et al. 1997) to 26% (Cairns et al. 1997) of the total biomass. Below-ground biomass accumulation is linked to the dynamics of above-ground biomass. The greatest proportion of root biomass occurs in the top 30 cm of the soil surface (Bohm 1979; Jackson et al. 1996). Revegetation of degraded land leads to continual accumulation of below-ground biomass whereas any disturbance to topsoil leads to loss of below-ground biomass.

Since below-ground biomass could account for 20–26% of the total biomass, it is important to estimate this pool for most carbon mitigation as well as other land-based projects. Estimation of stock changes in below-ground biomass is also necessary for greenhouse gas inventory at national level for different land-use categories such as forest lands, cropland and grassland. This chapter presents methods of estimating and monitoring below-ground biomass.

11.1 Below-Ground Biomass: Features and Broad Methods

Methods for measuring and monitoring above-ground biomass are relatively well established, in regular use and cost-effective; however, those for below-ground biomass are less developed and less frequently used in the field. Further, the methods for below-ground biomass for different land-use systems are not standardized (IPCC 2006). Live and dead roots are generally not distinguished and hence root biomass is reported as total of live and dead roots. The methods for estimating and monitoring below-ground biomass are listed below:

1. Excavation of roots
2. Monolith for deep roots
3. Soil core or pit for non-tree vegetation
4. Root to shoot ratio
5. Allometric equations

The choice of method depends on site conditions, vegetation type and the accuracy required, but in most carbon inventory projects root to shoot ratio and allometric equations are the most commonly used as will be explained in this chapter. Data on below-ground biomass are required for estimating and projecting total change in carbon stock for the following:

- Baseline scenario land-use systems
- Project scenario land-use systems

Such estimates are also required during project development and monitoring phases of a project cycle.

Project development phase Below-ground biomass is estimated and projected during the project development phase largely based on the default values of root to shoot ratio or allometric equations for tree biomass. Default values are also used for non-tree vegetation.

Project monitoring phase In project monitoring phase too, because of the large human effort and cost involved, below-ground biomass is normally estimated using the default root to shoot values or allometric equations. However, if suitable allometric equations or values of root to shoot ratio are not available for the location and the species, the below-ground biomass has to be measured physically, although it is a destructive method (Sections 11.2 and 11.3).

11.2 Excavation of Roots

If the below-ground biomass of trees is considered an important pool in a project, it may be necessary to measure the biomass. It may be possible to measure the below-ground biomass especially if there are only a few tree species in the project area or if the project area consists of a monoculture plantation. Since the measurement method is quite complex, requires large human effort and destroys the trees, it may be adopted only when no root biomass equations suitable for the species or the project location are available. This method is mostly used in tree-based systems and involves selecting the plots, excavating all roots, measuring their fresh weight, converting it to dry weight and extrapolating to per tree or per unit area (per hectare). The following steps could be adopted:

Step 1: Select and stratify land-use category or project activity (refer to Chapter 10).
Step 2: Select and locate plots in each stratum:

- Use the plots selected for shrub measurements within the tree quadrats (Chapter 10).
- Normally, select eight to 10 shrub plots for each stratum.
- If the aim is to estimate the root biomass of trees, select all the trees in the plots and number them for excavation.
- If the aim is to estimate root biomass on area basis, excavate the whole plot.

Step 3: Assemble the required material:

- Tape for measuring DBH and height
- Balance for weighing the shrub biomass
- Rope and pegs for marking plots
- GPS for locating sample plot boundaries

Step 4: Measure the trees, shrubs and herbs:

- Adopt the methods described in Chapter 10.
- Record DBH, height, species and other parameters for trees.
- Harvest and measure the weight of the shrub and herb biomass.

Step 5: Harvest all the trees to the ground level and number them.

Step 6: Weigh each tree thus harvested:

- Record the weight of the tree along with its DBH and height against its number.

Step 7: Excavate the roots of trees and other non-tree plants separately by digging and loosening the soil:

- Surface soil around the tree stem is excavated
  - If the aim is to estimate the root biomass of individual trees; in that case, mark the area around the trees for excavation until bulk of the roots belonging to that tree are included (the plot boundary need not be a restriction).
  - If the aim is to estimate the root biomass for the plot area, roots going beyond the plot boundary are excluded, although this may be difficult in practice.
- Since root biomass is concentrated in the top 30–50 cm layer, a minimum depth of 30 cm is necessary; for root biomass beyond that depth, refer to the monolith method (Section 11.3).
- Separate the roots of trees and non-tree biomass, although this may not be feasible in situations with mixed tree species and a mix of tree, shrub and herb plants.
- The soil along with roots may have to be washed in a sieve (mesh size of 2.5 or 5 mm) to separate the roots from the soil.
- If feasible, store the roots of each tree in a separate cloth bag after recording the number, DBH and weight of the tree.

Step 8: Measure the fresh weight of roots:

- For estimating root biomass of an area, it is desirable to estimate the weight of root biomass for each tree separately and pool them later.
- Pool all the non-tree biomass separately for each shrub plot.
- Take samples of root biomass (about 0.5 kg each) for each of the dominant tree species and for non-tree biomass separately for drying.
- Take fresh weight of samples for each tree species separately.

Step 9: Estimate the dry weight of the fresh root biomass by drying in the oven to a constant weight:

- Estimate dry weight separately for each dominant species and for the pooled non-tree biomass.

Step 10: Extrapolate the dry weight of the biomass to per hectare and stratum level:

- First, extrapolate according to individual tree and species at the plot level.
- Second, extrapolate for the whole plot.
- Third, extrapolate to per hectare.
- Finally, extrapolate to the stratum area from which the samples are taken.

The main disadvantages of the method are the large human effort, high cost, destruction of trees and disturbance of soil, leading to loss of carbon from the plots. However, if the measurement method is adopted, use the information collected to develop regression equations. The following relationships can be developed:

- Above-ground biomass equation based on DBH and height values of the trees
- Below-ground biomass equation using DBH and height of the trees
- Below-ground biomass equation using above-ground biomass values
- Root to shoot ratio

Parameters to be recorded for estimating root biomass are shown below (apart from information on name of location, land-use category, project activity, tree quadrat number, GPS reading, date of recording and name of investigator, as given in Chapter 10 for tree quadrats).

Tree species	Tree number	DBH (cm)	Height (m)	Fresh weight of shoot biomass (kg)	Fresh weight of root biomass (kg)	Dry weight of shoot biomass (kg)	Dry weight of root biomass (kg)

11.3 Monolith for Deep Roots

The monolith method is adopted for estimating root biomass below the depth of 30 cm and is mostly used in non-tree based land-use systems, such as grassland. The procedure involves cutting a monolith that is a large block of soil from a plot, separating the roots and weighing them. This method is used to obtain quantitative estimates of root biomass (FAO 2004). The broad steps in using the monolith method (Weaver and Darland 1949) are as follows:

Step 1: Select and stratify the land-use category or project activity.
Step 2: Dig a deep pit at the sample site ($1 \times 1 \times 1$ m).
Step 3: Make the wall of the pit smooth and vertical.
Step 4: Take a long shallow wooden or steel box (30 cm wide and 8 cm deep) without a top and lacking one of the four sides.

Step 5: Hold the box against any of the smoothened walls with its bottom away from the wall and the open side facing the sky but flush with the top of the pit.

Step 6: Press or hammer the box against the wall so that the three edges of the box leave three clear dents or grooves, two vertical and one horizontal, on the wall of the pit.

Step 7: Remove the box, take a knife and deepen the three grooves up to 10 cm, making a column of soil 10 cm broad.

Step 8: Fit the box around the column holding it in place and cut the column free from the other side with a knife or a spade and free the monolith from the surrounding soil.

Step 9: Wash the column of soil under running water to separate the roots embedded in the column.

Step 10: Measure the fresh weight and estimate the dry weight of the root biomass.

Step 11: Calculate the volume of the soil monolith extracted.

Step 12: Estimate the root biomass in grams in the volume of soil monolith extracted and extrapolate to per hectare to the depth of the monolith.

Among the disadvantages of the method are the facts that some roots may be lost during washing, that it is time-consuming and expensive and that excavating large monoliths often requires expensive machinery (MacDicken 1997). Further, the method can only be used with satisfying results in easily workable soils (Majdi 1996); it is not applicable to most land-use categories or project activities.

11.4 Soil Core or Pit for Non-Tree Vegetation

It is not feasible to estimate the root to shoot ratio or develop a root biomass equation linking above-ground biomass to root biomass in the case of non-tree vegetation such as herbs, shrubs and grasses; with such plants, the soil core method is the one most commonly used for the purpose. The broad steps in estimating the stock of below-ground biomass at a given point in time for a given land-use system by the soil core method are as follows:

Step 1: Select and stratify the land-use category and project activity.

Step 2: Select sampling plots for root biomass:

- Select the plots by following the procedure given for selecting shrub plots in Chapter 10.
- Select 8–10 shrub plots for each stratum and mark the centre of each plot for sampling root biomass.

Step 3: Assemble the material required:

- A soil core sampler (5–10 cm in diameter and 30 cm deep), metallic sieve for washing roots, balance for weighing

Step 4: Drive or insert the core sampler into the ground at the selected point and remove the soil along with roots.

Step 5: Separate the roots from soil by placing the soil samples on a sieve (mesh size of 2.5 or 5 mm) and wash the roots under running water.
Step 6: Collect all the roots and weigh them.
Step 7: Estimate the dry weight by oven-drying a sample of roots at 70°C to a constant weight (for at least 8 h).
Step 8: Estimate the dry weight of the roots for the volume of the core sampler (calculated from its diameter and depth) for all the sample plots.
Step 9: Extrapolate the root biomass to per plot and hectare, using dry weight of the root samples collected to the depth of the core.

11.5 Root to Shoot Ratio

There is a relationship between biomass in shoot and roots for a tree of a given species as well as for a given forest or plantation type. It is possible to estimate the biomass in roots based on data on above-ground biomass. Given the limitations of measuring below-ground biomass in the field, it is desirable to use indirect methods to obtain the value of below-ground biomass. A comparative review by Cairns et al. (1997) includes more than 160 studies covering native tropical, temperate and boreal forests that reported both below-ground biomass and above-ground biomass. The below-ground (root) biomass to average above-ground (shoot) biomass ratio developed based on these studies was 0.26 with a range of 0.18–0.30. The ratios did not vary significantly with latitude (tropical, temperate or boreal), soil texture (fine, medium or coarse) or tree type (angiosperms or gymnosperms).

Specific root to shoot ratios could also be developed for the project location and species, if necessary, by adopting the root excavation method described in Section 11.2. The key steps are as follows (for details, refer to Section 11.2):

Step 1: Select the tree species for root biomass measurement from the strata and sample plots for measurement. Select trees of different DBH or height; the selection need not be confined to the sample plots.
Step 2: Harvest a selected number of trees for excavation; about 30 trees of different girth sizes may be adequate.
Step 3: Excavate the roots up to 30–50 cm depth, clean the roots of soil and take their fresh weight.
Step 4: Measure the tree DBH, height and shoot (above-ground biomass) weight (refer to Section 11.2).
Step 5: Measure the fresh weight and estimate dry weight of the root biomass excavated.
Step 6: Estimate the weight of shoot biomass using the DBH (and height)-based biomass equations (Refer to Chapter 17).
Step 7: Calculate the root to shoot biomass ratio from the estimated weights.

The development of root to shoot ratio involves a large human effort and cost. The root to shoot (R) ratios for tropical forests (Mokany et al. 2006) are as follows:

- Tropical moist deciduous forest: R = 0.20 (0.09–25) for forests with above-ground biomass less than 125 t/ha
- Tropical dry forest: R = 0.28 (0.27–0.28) for forests with above-ground biomass greater than 20 t/ha

11.6 Allometric Equations

In addition to root to shoot ratio, allometric equations have been developed linking above-ground biomass to below-ground biomass. Such equations could be developed for an individual tree species or for a plantation or a forest type. Allometric equations for broad forest types, as developed by Cairns et al. (1997), are illustrated in Chapter 17.

It is important to note that allometric equations are developed based on observations from native forests and thus may not be applicable to plantations or other vegetation types. However, in the absence of allometric equations specific to location, forest type and species, the broader tropical or temperate or boreal equations could be used. For example, root biomass (Y) for tropical forests (dry tonnes/hectare) could be calculated using the following equation (Cairns et al. 1997).

$$Y = \text{Exp}\ [-1.0587 + 0.8836 * \text{LN(AGB}]$$

where LN = natural logarithm, AGB = above-ground biomass (dry tonnes/hectare).

Estimation of below-ground biomass using the allometric function requires the estimated value of above-ground biomass using methods given in Chapters 10 and 17 and of below-ground biomass using methods given in Sections 11.2–11.5. These equations provide below-ground biomass values in tonnes of dry biomass per hectare. The method described for developing root to shoot ratio in Section 11.5 could be adopted for developing the allometric equations.

If region- and species-specific root to shoot ratios are available, these could be used for carbon inventory; if not, the general equations developed by Cairn et al. (1997) may be used.

11.7 Long-Term Monitoring of Below-Ground Biomass

Below-ground biomass accumulates over decades and centuries. The below-ground biomass stock is linked to the above-ground biomass stock. Changes in stocks of above-ground biomass will be monitored over long periods in all projects using the methods given in Chapter 10. Periodical harvesting and extraction of below-ground biomass for estimation is not feasible in most situations. Thus, the practical approach is to monitor and estimate the above-ground biomass stock and use the root to shoot ratio or allometric equations to calculate the below-ground biomass stock. Thus, there is no need for long-term monitoring of below-ground biomass stocks.

11.8 Conclusions

In forests and plantations, below-ground biomass could account for about a quarter of the total biomass. In grassland and cropland, bulk of the below-ground biomass is part of the annual cycle. Estimation of below-ground biomass is important largely for tree-based land-use categories or projects such as forests and plantations, especially for cropland, grassland and degraded forest land converted to forests or plantations, where root biomass accumulates. Measurement and estimation of root biomass is complex and expensive and leads to loss of vegetation and disturbance to topsoil. Therefore, measurement should be resorted to only if specifically needed. Below-ground biomass is estimated for most carbon mitigation projects and national GHG inventory using root to shoot ratio or allometric equations, which require estimates of above-ground biomass. There is high correlation between above-ground biomass and below-ground biomass, and the root to shoot ratio varies within a narrow range. Thus, the default root to shoot ratios or allometric equations could be used in carbon inventory programmes.

Chapter 12
Methods for Dead Organic Matter: Deadwood and Litter

Dead organic matter consists of deadwood and litter. Stems and branches of deadwood 10 cm or larger in diameter form the deadwood pool and those smaller than that constitute litter (see Chapter 4 for a definition). Inclusion of dead organic matter pool makes the estimated changes in total carbon stock more accurate. Most of the biomass not harvested or burnt is added to the deadwood, litter and soil carbon pools. The dynamics of dead organic matter vary with the type of forest or plantation as well as with the purpose behind protecting a forest or raising a new forest. In fuelwood plantations or community forestry projects, the woody part of the dead organic matter is likely to be removed and used as fuelwood. However, in the case of avoided deforestation projects involving protection of forests, dead organic matter accumulates on the forest floor. Further, land-use change, particularly from forests and plantations to other land uses such as cropland or grassland, leads to complete loss of dead organic matter. Dead organic matter is not likely to be a dominant carbon pool for grassland reclamation, agroforestry and cropland management projects: it may account for about 10% of total carbon stocks in forests and tree plantations (Chapter 4) but may be practically absent in other land-use categories.

$$DOM = DW + LT$$

where DOM = dead organic matter, DW = deadwood and LT = litter.

The relevance of dead organic matter to project development and monitoring phases of land-based projects is as follows:

- *Project development phase* Dead organic matter is not relevant to baseline scenario, except for avoided deforestation or conversion of forest land to other land uses. Dead organic matter pool is either ignored for the project scenario or estimated using default values during the project development phase.
- *Project monitoring phase* During the project monitoring phase, dead organic matter is likely to be an important pool for carbon mitigation projects such as avoided deforestation, conversion of forest land to other uses and afforestation and reforestation projects. Thus, dead organic matter could be periodically measured and estimated during the project monitoring phase. This pool is

relevant neither to commercial roundwood production programmes nor to non-tree land-based projects on grassland and cropland.

Dead organic matter pools can be estimated by "Gain–Loss" or "Stock-Difference" methods. These two methods are explained in IPCC (2003, 2006) and in Chapter 9. The methods and procedures for measuring and estimating deadwood and litter are explained in this chapter.

12.1 Deadwood

Deadwood includes all non-living woody biomass not contained in the litter, either standing, lying on the ground, or in the soil. Deadwood includes wood lying on the land surface, dead roots and stumps 10 cm or larger in diameter. Methods for estimating deadwood are considered at two levels (Pearson et al. 2005b), namely standing deadwood and fallen deadwood.

12.1.1 Standing Deadwood

Standing deadwood usually includes trees that are dead but not yet fallen to the ground and thus are part of the vegetation. The trees may have died because of disease or physical damage. Standing deadwood could be a key carbon pool in older forests and plantations but unlikely to be so in young plantations, cropland and grassland development projects. The method adopted for estimating standing deadwood is identical to that used for estimating above-ground biomass. The key steps in estimation of the stock of deadwood at a given time are as follows (based on steps given in Chapter 10):

Step 1: Select and stratify the land-use category or project activity for which the deadwood has to be estimated.

Step 2: Decide on the sampling method including sample size, number of sample plots and sampling design, used for estimating the above-ground tree biomass of trees.

Step 3: Select the sample plots identified for above-ground biomass and use the same plots.

Step 4: Assemble the material required for field study, namely:

- Measuring tape for recording diameter and height, rope and pegs for marking the plot, a balance for weighing fresh wood samples, a cotton bag for storing samples, and a knife for cutting litter

Step 5: Identify the parameters to be measured and recorded:

- DBH and height
- The status of the dead and standing trees, based on expert judgement

  - Tree with crown, branches and twigs but without leaves
  - Tree without crown and branches
  - Tree stump (with a short stem)

- Wood density

  - Measure fresh weight, dry weight and volume of the sample wood block

Step 6: Record these parameters together with those for live above-ground biomass:

- A suggested format for recording field data on dead and standing trees is shown below (in addition to location, land-use category or project activity, plot size and number, GPS reading, date and name of the investigator).

Tree Quqdrat number	Tree species	Tree number	Status*	DBH(cm)	Height(m)

*Tree with crown, branches and twigs: tree without crown and branches; tree stump

Step 7:Calculate the biomass using methods given in Chapter 17. The method for calculating the biomass stock of dead and standing trees is identical to that used for estimating above-ground biomass.

12.1.2 Fallen Deadwood

Fallen deadwood occurs in forests and older plantations and could be the result of natural death or of strong winds, disease or pest attack and felling. In many project areas, fallen deadwood may be removed by the local communities for fuelwood and other uses. The sampling method could be identical to that used for estimating above-ground biomass. The following are the key steps in measuring fallen deadwood:

Step 1: Select and stratify the land-use category or project activity for which the fallen deadwood has to be estimated (Chapter 10).

Step 2: Decide on the sampling method including sample size, number of plots, and sampling design used for estimating above-ground biomass of trees.

Step 3: Select the sample plots identified for above-ground biomass and use the same plots.

Step 4: Assemble material required for field study, namely:

- Measuring tape for recording diameter and length and rope and pegs for marking the plot, and balance for weighing deadwood

Step 5: Identify the parameters for measuring fallen deadwood:

- Measure the diameter or girth at both ends and midpoint, and length of the fallen tree or branch

  - If more than half of the fallen wooden log or branch is inside the plot boundary, include it as part of the plot for measurement

- Record status of the fallen wood, based on expert judgement

  - Good physical condition (not rotten or decomposing)
  - Cavity formed in the middle due to rotting or decomposition

- A suggested format for recording field data is shown below (in addition to the location, land-use category or project activity, plot size and number, date and name of the investigator)

Tree quadrat number	Tree species[a]	Log/branch number	Status[b]	Diameter(cm)			Length(m)	Weight of the stem/ branch (kg)
				Tip	Mid	Bottom		

[a]Identify the species of dead and fallen wood log or branch, if possible
[b]Good physical condition (not rotten or decomposing); cavity formed in the middle

- If cavity exists in the middle of the fallen stem, measure the diameter of the cavity (and exclude it from the volume calculation of the fallen deadwood)
- If the fallen stem is not very long or large and if feasible, measure the weight

  - Using a spring balance by tying a rope to the fallen deadwood and lifting it

- Measure the weight of all the fallen deadwood logs and if weighing is not possible, measure the length and the diameter
- Record the observations in the suggested format and enter the data in a database
- Take a sample of deadwood for density estimation

  - Measure fresh weight; estimate the dry weight and volume of the sample wood block

- Calculate the biomass using the methods given in Chapter 17

12.2 Litter Biomass

Litter includes all non-living biomass other than deadwood (normally less than 10 cm in diameter lying dead in various states of decomposition above the mineral soil. Litter, which includes woody and non-woody components, consists of plant parts that fall to the ground as part of the annual cycle, as a result of pest attack, physical damage such as that due to wind or lopping of branches and leaves, and could be further categorized into coarse woody litter (diameter greater than 6 mm), fine woody litter (diameter 6 mm or smaller) and non-woody litter (leaves and reproductive parts). Litter biomass, lying on the top of the mineral soil and on the floor of forests or plantations, could be a key pool only for older forests and plantations.

Litter may not be a key pool in barren or degraded lands, particularly in the baseline scenario.

Refer to Chapter 4 on the importance of measuring litter pool for different project types as well as for the frequency of measurement. Litter biomass can be measured using two methods:

(i) Annual litter production
(ii) Litter stock change method

12.2.1 *Annual Litter Production Method*

Litter production is measured to assess the annual woody and non-woody litter fall as well as the turnover rates expressed as dry tonnes/hectare/year. Estimating annual litter biomass production is a complex task and involves fixing litter traps in all the sample plots and collecting and weighing the litter every month. This requires protecting litter traps and preventing removal of litter from litter traps in field situations. The method involves significant human effort. The following steps could be adopted for estimating annual litter production:

Step 1: Select and stratify the land-use category or project activity.
Step 2: Select the plots earmarked for measuring shrub biomass (Chapter 10):

- Normally 8–10 plots per stratum, each measuring 5 × 5 m

Step 3: Assemble material required for field study, namely:

- XRope and pegs for marking the plot, litter trays, a balance for weighing litter, cotton bags for collecting sample for drying, a knife for cutting litter

Step 4: Fabricate square litter trays made of 1.5 mm wire mesh with a wooden frame:

- Dimensions = 1 × 1 × 0.1 m (depth) with the wooden frame
- Number = 8–10 per stratum

Step 5: Install the litter trays randomly in the shrub plots by fixing the trays about 15 cm above the ground to ensure that water does not accumulate in them and that they are not easily removable or damaged.
Step 6: Collect the litter fallen into the trap once in a month on a fixed date. Remove deadwood (>10 cm diameter) if any.
Step 7: Record the fresh weight of the litter:

- Separate the woody and non-woody litter
- Weigh each component separately for each trap
- Take about 0.5 kg of fresh litter for dry matter estimation

Step 8: Repeat the field collection and measurement of litter once in every month for 12 months.

Step 9: Calculate and tabulate dry weight of woody and non-woody litter on monthly basis.

Step 10: Calculate the per hectare annual litter production using the following procedure:

Pool the weights of woody and non-woody litter from all the plots in a stratum
Convert the fresh weight to dry weight on a monthly basis
Add the total dry weight from all the plots for each of the 12 months
Extrapolate total dry annual litter fall from sample plots to dry tonnes/hectare/year

12.2.1.1 Merits and Demerits of Measuring Annual Litter Production

Merits: It is possible to obtain reliable estimates of annual litter production. The method provides separate estimates for woody and non-woody litter and even by species, if necessary.

Demerits: Visiting the site every month for a year is expensive. Protection of litter trap-trays in the field is difficult since they may be damaged by grazing or wild animals or stolen. Some of the litter could be blown away by wind.

12.2.2 Litter Stock Change Method

The litter stock change method estimates the stock of litter biomass at a given time. The litter stock could be measured at two points a year apart or even 5 years apart. The method involves collection and weighing of the litter in the selected sample plot and the following steps:

Step 1: Select and stratify the land-use category or project activity for which the litter biomass has to be estimated (Chapter 10).

Step 2: Decide on the sampling method, including sample size, number of plots, and sampling design used for estimating shrub biomass (usually 8–10 shrub plots of 5 × 5 m each).

Step 3: Assemble the material required for measurement, namely:

- A spring balance, measuring tape, pegs and ropes to mark the four corners, a heavy-duty cutting tool for cutting wood and a knife for cutting litter

Step 4: Mark the four corners of each shrub plot and also record the GPS readings.

Step 5: Sweep and collect all the litter from the sample shrub plots and separate it into woody and non-woody litter. In collecting litter near the boundary, include any piece of litter so long as at least half its length is within the sample plot.

Step 6: Estimate the fresh weight of the litter in the field for each sample plot.

Step 7: Estimate the dry weight by collecting a sample of about 0.5 kg and drying it to a constant weight in an oven.

Step 8: Extrapolate the sample plot value of dry litter to per hectare stock of the litter.

This method estimates the carbon stock in the litter pool at a given point in time; changes in the stock can be estimated by repeating the measurements after 1 year.

12.2.2.1 Merits and Demerits of Measuring Stock Change

Merits: Stock change is a simple and cost-effective method requiring minimal field visits, measurements and materials.

Demerits: The litter biomass may be removed for fuelwood or other purposes or lost by being blown away, losses that are not captured by this method. Estimation of litter using the annual litter production method is complex and expensive, and it is important for project managers to choose between measuring litter production and monitoring litter stock changes.

12.3 Long-Term Monitoring of Deadwood and Litter

Deadwood and litter in land-use systems such as forests and plantations accumulate over long periods if the biomass is not removed. These two pools could be monitored by adopting the "Stock-Difference" approach and the permanent plot method.

Deadwood Select the permanent sample plots marked for long-term monitoring of above-ground tree biomass (Chapter 10) and adopt the method described for standing and fallen deadwood in Section 12.1 for long-term monitoring of changes in stocks of deadwood.

Litter Select the permanent sample plots marked for shrubs for long-term monitoring of litter using "Stock-Difference" approach. Adopt the method described in Section 12.2.2 for long-term monitoring of litter stock changes.

12.4 Conclusions

Dead organic matter consists of standing and fallen deadwood and litter biomass. These pools are likely to be significant for forests and plantations but not for other land-use categories or project activities. Both the deadwood pools can be measured easily by using the methods described in this chapter, simultaneously with measurement of above-ground biomass pool (Chapter 10), with very little additional cost or human effort. Similarly litter biomass could be estimated using the stock change measurement method, along with above-ground biomass (Chapter 10), with very little additional cost or human effort. However, estimating annual litter production is complex and time consuming, and expert judgment is needed in deciding whether dead organic matter should be measured, especially since it accounts for only about 10% of the total carbon stocks in forests.

Chapter 13
Methods for Estimating Soil Organic Carbon

Soil is the largest reservoirs of carbon, accounting for 2011 GtC, or 81% of the total carbon in the terrestrial biosphere (WBGU 1998). Flow of carbon between soil and the atmosphere is a continuous process, highly influenced by land use and management (Paustian et al. 1997). Organic carbon stored in soil is an important carbon pool for many land-use systems and projects, and even for national greenhouse gas inventories of different land-use categories. "Soil organic carbon" is also often referred to as "soil organic matter". Soil organic matter includes the whole non-mineral fraction of soil ranging from decayed plant and animal matter to brown to black material that bears no trace of the original anatomical structure of the material and is normally defined as "soil humus". Soil organic matter also includes living and dead microbial tissue, compounds synthesized by microorganisms and derivatives of these materials produced as a result of microbial decay. Soil organic carbon as defined by IPCC (2006) comprises "organic carbon in mineral soils to a specific depth chosen, also including live and dead fine roots within the soil". Although both organic and inorganic forms of carbon are found in soil, land use and management typically has a larger impact on the stocks of organic carbon. Therefore, this chapter focuses on soil organic carbon. Further, soil organic carbon is relevant to both mineral and organic soils. Organic soils contain a minimum of 12–20% organic matter by mass and are found under poorly drained conditions of wetlands (Brady and Weil 1999). All other soils are classified as mineral soils, which typically have relatively low amounts of organic matter. Mineral soils dominate most ecosystems except wetlands and are the focus of this chapter.

Stocks of organic carbon in soil vary with land-use systems. The share of soil organic carbon in the total carbon stock may vary from 50% to 84% in forests to 97% in grasslands (Bolin and Sukumar 2000). Soil organic carbon is likely to be a dominant pool in all land-use systems without trees such as grassland and cropland. Carbon stocks in soils are fairly stable under undisturbed conditions, such as natural forests or grasslands. Land-use change involving disturbance to topsoil leads to oxidation of organic matter and rapid loss of soil organic carbon. The concentration of organic carbon in soil is the highest in topsoil. Soil carbon dynamics is normally restricted to the top 15–45 cm, which is the zone of maximum microbial activity. Soil organic carbon is normally estimated to a depth of 0–30 cm since most of it is present in the top layers and root activity is also concentrated in this horizon.

Change in the stocks of soil organic carbon in mineral soils is estimated using the following formula:

$$\Delta SC = \frac{(SC_{t_2} - SC_{t_1})}{(t_2 - t_1)}$$

where:

ΔSC = annual change in carbon stock in mineral soil, tonnes C/year
SC_{t_1} = soil organic carbon stock at the beginning of the period t_1, tonnes C/ha
SC_{t_2} = soil organic carbon stock at period t_2, 5 or 10 years, tonnes C/ha.

13.1 Soil Carbon Inventory for Land-Use Projects and Greenhouse Gas Inventory

13.1.1 Soil Carbon Inventory for Mitigation Projects

An inventory of soil organic carbon (SOC) pool is necessary for carbon mitigation projects (such as afforestation, reforestation and agroforestry) as well as all land reclamation projects such as grassland reclamation, shelterbelts and watershed projects. Estimates of SOC are required for the following scenarios:

- *Baseline scenario* where SOC is estimated prior to initiation of the project activities as well as in the control plots by simulating baseline scenario conditions during the project monitoring phase if SOC is expected to change under the baseline scenario.
- *Project scenario* involving periodic estimation of SOC in land-use systems subjected to project activities, classified into homogeneous strata.

Most carbon mitigation as well as land reclamation projects increase the stocks of SOC. A carbon inventory should provide estimates and projections of SOC during the following phases:

- *Project development phase*, where SOC is estimated for all land-use systems and strata (see Section 10.3 for strata definition) before the implementation of project activities and projected into the future.
- *Project monitoring phase*, where SOC is measured and estimated periodically for land-use systems in which project activities are implemented, as well as for the baseline scenario land-use systems.

Soil organic carbon is routinely estimated in most land-based projects as an indicator of the impact of project activities on stock of soil carbon, soil fertility, moisture-holding capacity or soil erosion. Researchers and project managers in forestry, agriculture, grassland development and conservation projects estimate the stocks of SOC and changes in those stocks. Methods for estimating SOC are well documented and extensively used in all projects (IPCC 2003, 2006; Mac Dicken 1997; Hairiah et al. 2001).

13.1.2 Soil Carbon Inventory for National Greenhouse Gas Inventory

Estimates of emissions and removals of SOC or stock changes in mineral and organic soils in land-use categories and subcategories are required for national greenhouse gas inventories. Stocks of SOC in different land-use categories need to be estimated for a selected inventory year using carbon stocks estimated at two points in time separated by several years because measurement may not be feasible over a period of 1 year. The methods described in this chapter could be adopted for generating emission and/or sequestration factors relevant to SOC in estimating national greenhouse gas inventories (Chapter 16).

13.2 Methods for Inventory of Soil Organic Carbon

Several methods are available and in use for estimating SOC, ranging from simple laboratory estimations to diffuse reflectance spectroscopy.

- Wet digestion or titrimetric determination (Walkley and Black method)
- Colorimetry
- Direct estimation of organic matter by loss-on-ignition
- CHN analyser
- Diffuse reflectance spectroscopy
- Modelling

The most common method used in the field is the wet digestion or titrimetric determination method, which is also a cost-effective method. The carbon, hydrogen and nitrogen (CHN) analyser, although very accurate, is rarely used in field studies because the instrument is expensive. Diffused reflectance spectroscopy is also expensive and yet to be widely used in field. Modelling is limited by the availability of models and data to represent local conditions. Remote sensing method can be used only for large projects, but still needs modelling and validation by data obtained from other methods.

(i) *Wet digestion or titrimetric determination* Wet digestion involves a rapid titration procedure to estimate the organic carbon content of soil (Kalara and Maynard 1991). *Principle* Organic matter is oxidized with a mixture of $K_2Cr_2O_7$ and H_2SO_4. Unused $K_2Cr_2O_7$ is back-titrated with ferrous ammonium sulphate (FAS). Organic carbon in the soil is oxidized to CO_2.

Material A burette, pipette, 500 ml conical flask, measuring cylinder and analytical balance.

Reagents

- 1N $K_2Cr_2O_7$solution: dissolve 49.04 g of $K_2Cr_2O_7$ in minimum amount of distilled water and make up the final volume to 1 l
- 0.5 N ferrous ammonium sulphate (FAS) or Mohr's salt: dissolve 392 g FAS in distilled water. Add 15 ml of concentrated H_2SO_4 and make up the volume to 2 l with distilled water
- Diphenylamine indicator: dissolve 0.5 g of diphenylamine in a mixture of 100 ml of concentrated H_2SO_4 and 20 ml of distilled water
- Concentrated H_2SO_4 containing 1.25% Ag_2SO_4 (silver sulphate): if the soil is free of chlorides, use of Ag_2SO_4 can be avoided
- Sodium fluoride (NaF) or orthophosphoric acid 85%

Procedure

1. Weigh 0.5 g of powdered and sieved (2 mm) soil into a 500 ml conical flask
2. Add 10 ml of 1 N $K_2Cr_2O_7$ solution and shake to mix
3. Add 20 ml of concentrated H_2SO_4 from the sides of the flask
4. Keep the contents of the flask undisturbed for 30 min
5. Add 3 g NaF or 10 ml of H_3PO_4 and 100 ml of distilled water and shake vigorously
6. Add 10 drops of diphenylamine indicator, which turns the solution violet
7. Titrate against 0.5 N FAS solution until the colour changes from violet to bright green and note the volume of solution used
8. Carry out a blank titration in a similar manner without the soil.

Calculations

- Weight of the sample = Sg
- Volume of FAS used in blank = Xg
- Volume of FAS used to oxidize SOC = Yg
- Normality of FAS = N
- Volume of 1 N $K_2Cr_2O_7$ used for the oxidation of carbon = (X−Y)/2
- 1 ml of 1 N $K_2Cr_2O_7$ = 0.003 g SOC
- *Percentage of organic carbon in the soil* = [(X−Y)/2 × 0.003 × 100]/S

Conclusion This is the most commonly adopted method in most laboratories with minimal facilities, as it does not require sophisticated equipment. This is also the least expensive method and the results are reasonably accurate. If CHN analyser is available, it is desirable to compare the results from the wet digestion method with those from the CHN analyser for validation and adopt a correction factor if necessary.

(ii) *Direct estimation of organic matter by loss-on-ignition*

Principle Organic matter is oxidized by heating at 375 °C and estimated by weight loss.
Material A muffle furnace, porcelain crucibles, a desiccator.

Procedure

1. Heat porcelain crucibles for 1 h at 375 °C.
2. Cool in open to about 150 °C. Place in a desiccator, cool for 30 min and weigh.
3. Weigh about 5 g oven-dried (to the nearest milligram) sample, passed through a 2 mm sieve, into each crucible.
4. Place the crucibles containing the samples in a muffle furnace at room temperature. Heat slowly (increase temperature by about 5 °C every minute) to 375 °C ± 5 °C.
5. Maintain at 375 °C ± 5 °C for 16 h.
6. Turn furnace off and let temperature drop to about 150 °C.
7. Remove crucibles and place in desiccator for 30 min. Weigh to the nearest milligram.

Calculation
Loss on ignition (%)

$$\text{Organic matter \%} = \frac{[\text{weight of oven-dried sample (g)} - \text{weight of sample after ignition (g)}] \times 100}{\text{weight of oven-dried sample (g)}}$$

Merits and demerits The loss-on-ignition method of estimating organic matter is sufficiently accurate for most descriptive purposes. The method is most suitable for well-aerated samples (e.g. sandy and peat soils) with low clay mineral and inert carbon (charcoal) content.

However, the method is not suitable for calcareous soils. The procedure is prone to error as the weight loss may include carbon from carbonates and water and from hydroxyl groups from clay. Error is also caused by combustion of inert carbon compounds and volatilization of substances other than organic material. There is incomplete oxidation of carbonaceous materials in some soils at 375 °C.

(iii) *Carbon, hydrogen, nitrogen (CHN) analyser* Total organic carbon is a measure of the total amount of non-volatile, volatile, partially volatile and particulate organic compounds in a sample. Total organic carbon is independent of the oxidation state of the organic compounds and is not a measure of the organically bound and inorganic elements that can contribute to tests of biochemical and chemical oxygen demand.

Principle A CHN analyser analyses the carbon in solids based on the Dumas concept (Macko 1981) using helium and oxygen gases. The carbon in the sample is heated and carbon is oxidized (CO_2) by oxygen in the presence of helium. The CO_2 evolved is directly proportional to the content of carbon in the sample. The carbon evolved is detected by a CO_2 detector. Results are represented as percentage.

Materials

- Induction furnace
  - Leco WR-12, Dohrmann DC-50, Coleman CHN analyser, Perkin Elmer 240 elemental analyser, Carlo-Erba 1106

- Analytical balance: 0.1 mg accuracy
- Desiccator
- Combustion boats
- 10% hydrochloric acid
- Cupric oxide fines (or equivalent material)
- Benzoic acid or other carbon source as a standard.

Equipment preparation

- Clean combustion boats by placing them in the induction furnace at 950 °C. After being cleaned, combustion boats should not be touched with bare hands
- Cool boats to room temperature in a desiccator
- Weigh each boat to the nearest 0.1 mg.

Sample preparation

- If samples are frozen, allow samples to warm to room temperature
- Homogenize each sample mechanically
- Transfer a representative aliquot (5–10 g) to a clean container.

Collection and storage Samples can be collected in glass or plastic containers. The recommended total weight of the soil sample, consisting of multiple aliquots, is 25 g. If unrepresentative material is to be removed from the sample, it should be removed in the field and a note to that effect made on the field data sheet. Samples should be stored frozen and can be held for up to 6 months under such storage. Excessive temperatures should not be used to thaw the samples.

Laboratory procedures

Step 1: Dry each sample to constant weight at 70 °C. The drying temperature is relatively low to minimize loss of volatile organic compounds.
Step 2: Cool the dried sample to room temperature in a desiccator.
Step 3: Grind the sample using a mortar and pestle to break up aggregates.
Step 4: Transfer a representative aliquot (0.2–0.5 g) to a clean, pre-weighed combustion boat.
Step 5: Determine sample weight to the nearest 0.1 mg
Step 6: Add several drops of hydrochloric acid to the dried sample to remove carbonates. Wait until the effervescing is complete and add more acid. Continue this process until the incremental addition of acid causes no further effervescence. Do not add too much acid at one time as this may cause loss of sample due to frothing.
Step 7: Dry the acid-treated sample to constant weight at 70 °C.
Step 8: Cool to room temperature in a desiccator.
Step 9: Add previously ashed cupric oxide fines or equivalent material (e.g. alumina oxide) to the sample in the combustion boat.
Step 10: Weigh the ascarite tube (A) before combustion.
Step 11: Combust the sample in an induction furnace at a minimum temperature of 950 ± 10 °C and weigh the ascarite tube.

Calculations If an ascarite-filled tube is used to capture CO_2, the carbon content of the sample can be calculated as follows:

$$\text{Percent carbon} = A\ (0.2729)\ (100)/B$$

where:

A = the weight (g) of CO_2 determined by weighing the ascarite tube before and after combustion
B = dry weight (g) of the unacidified sample in the combustion boat
0.2729 = the ratio of the molecular weight of carbon to the molecular weight of carbon dioxide.

A silica-gel trap should be placed near the inlet end of the ascarite tube to trap any moisture driven off during combustion. Additional silica gel should be placed at the exit end of the ascarite tube to trap any water that may be formed by the reaction between the trapped CO_2 and the NaOH in the ascarite.

If an elemental analyser is used, the amount of CO_2 will be measured by a thermal conductivity detector. The instrument should be calibrated daily using an empty boat blank as the zero point and at least two standards. The standards should bracket the expected range of carbon concentrations in the samples.

Conclusion A CHN analyser is a very reliable method, but the instrument is expensive to acquire and maintain. It can analyse a large number of samples at one time efficiently and effectively. This method is largely used for experimental studies and yet to find application in field projects.

(iv) *Diffuse reflectance spectroscopy* Diffuse reflectance spectroscopy (DRS) is a technology for non-destructive characterization of the composition of materials based on the interaction of visible–infrared light (electromagnetic energy) with matter. The method has the potential to increase efficiencies and reduce costs in both large-area applications (e.g. soil survey, watershed management, soil quality analysis) and site-specific management requirements. In particular, the ability to rapidly characterize a large numbers of samples using DRS opens up new opportunities in predicting and interpreting soil properties.

Principle Samples are illuminated with an artificial light source and the diffuse light reflected by the sample is collected and channelled through fibre optic cables to arrays of light detectors. The relative reflectance in each waveband comprises the reflectance spectrum for a sample, which is displayed and stored on a computer.

Procedure All soil spectral reflectance measurements are obtained using a Field spec ProFR spectroradiometer. The raw data must be processed before it can be used to predict soil properties. Multiple scans per sample allow the derivatives for a sample to be averaged and used in predicting soil properties.

Field sampling procedure

- Collect 20 soil cores (3 cm in diameter) to a depth of 1 m
- Mark each core location using a GPS so that the sample locations can be relocated in the future

- Dry and crush samples and pass through a 2 mm sieve

Laboratory procedure

- Pack air-dried soil samples, ground fine enough to pass through a 2 mm sieve, in polystyrene Petri dishes 55 mm in diameter and 12 mm deep
- Heap the Petri dishes with soil and scrape off excess soil using a blade to ensure a flat surface flush with the top of the dish
- Illuminate samples from above with two tungsten quartz halogen filament lamps in housings with aluminium reflectors
- Record the diffuse reflectance spectra of the samples using a FieldSpec FR spectroradiometer at wavelengths from 0.35 to 2.5 nm with a spectral sampling interval of 1 nm
- Record the average of ten spectra (the manufacturer's default value) at each position to minimize instrument noise
- Before reading each sample, record ten white reference spectra using calibrated spectralon placed at the same distance from the fibre optic as the soil sample
- Express reflectance readings for each wavelength band relative to the average white reference readings

Merits and demerits The main merits of DRS are higher precision and accuracy, repeatability and speed compared to conventional methods of soil analysis. With this method, a single operator can comfortably scan several hundred samples a day.

The main limitations of DRS include the need to build calibration libraries for a given population of soils for all the soil properties of interest and the complexity of the data analysis that this involves. Development of global calibration libraries in centralized laboratory facilities and software development for automated data analysis could help to reduce these limitations. Also, the method needs expensive instrumentation and trained staff.

Conclusion Further research and commercial development may lead to cheaper and more portable spectrometers, coupled with more flexible software and easier calibration methods. The technology should be increasingly used in a wide range of soil studies and surveys, and spectrometers are likely to become standard equipment in soil laboratories.

13.3 Broad Procedure for Soil Carbon Inventory

Making an inventory of soil carbon involves estimating the quantity of organic carbon present in the soil of a given land-use category or project activity at a given depth. Soil organic carbon is less frequently estimated compared to above-ground biomass, since the annual rate of change is low. A soil carbon inventory involves

- Estimation of bulk density of the soil at the specified depth
- Estimation of the concentration of organic carbon content in the soil sample
- Conversion of organic carbon content to tonnes of carbon per unit area (tC/ha) for a given depth of soil, using the bulk density.

The broad steps involved in inventory of soil carbon are as follows:

Step 1: Select the land-use category and project activities, stratify the area and demarcate project boundary according to the strata defined
Step 2: Determine the frequency of measurement
Step 3: Select the method for estimating

- Bulk density
- Soil organic carbon content

Step 4: Select the sampling technique
Step 5: Prepare for fieldwork
Step 6: Locate sampling points in the field
Step 7: Collect soil samples for laboratory analysis
Step 8: Measure bulk density parameters in the field
Step 9: Analyse the soil samples in the laboratory
Step 10: Enter field data and laboratory results into the database
Step 11: Calculate the quantity of soil organic carbon (tC/ha)

Determination of soil organic carbon requires access to laboratory facilities. It is important to decide whether soil organic carbon is a key pool and if this pool is likely to be impacted by the project activities and choose the frequency of estimation.

13.3.1 Selection of Land-Use Category Or Project Activities, Stratification of the Area and Demarcation of Project Boundary

Estimates of soil carbon are required at the beginning of the project, before implementing the project activities and periodically after the implementation of project activities during the monitoring phase. The procedure presented here is for a project or a set of project activities and could be adopted for land-use categories for GHG inventory also. The key steps involved are as follows:

(i) *Selection of project area* Identify and locate the project area on a map. The procedure for selection of project area is described in Chapters 8 and 10. This would involve all the land area that is likely to be subjected to project activities.
(ii) *Stratification of project area* Adopt the method described in Chapter 10 (Section 10.3) for stratification of project area based on physical, biological and management factors. It is better to adopt the same stratification procedure used for estimation of above-ground biomass, since soil organic carbon is estimated and added to biomass carbon pools, particularly the above-ground biomass pool.
(iii) *Definition of project boundary* The project boundary is defined and methods for demarcating it are described in Chapter 8. Adopt the same project boundary as that demarcated for above-ground biomass.
(iv) Preparation of maps Prepare maps depicting the project area, project activities, strata adopted and the project boundary. These maps are required during

- *Project development phase* for sampling prior to implementation of project activities.
- *Project monitoring phase* for locating sampling plots for periodic measurements.

The map prepared using a grid or a geo-referencing system will be necessary for locating the soil sampling points along with those for sampling the above-ground biomass during both the above phases.

13.3.2 Determination of the Frequency of Measurement

Soil carbon stocks in undisturbed soil are fairly stable; however, land-use change or any disturbance to topsoil leads to loss of carbon stocks. Conversion of grasslands and forests to other uses involving soil disturbance leads to loss of SOC. Besides land-use change, management practices can have a significant impact on stocks of SOC, particularly in cropping systems and grasslands. The frequency of measuring soil carbon stocks varies with land-use system, project activity and management system, as given in Chapter 4. The frequency has implications for selection of the method for preparing a carbon inventory method and for costs and can vary from once a year for activities involving land-use change to once in 5 years in majority of the projects. Using guidance from Chapter 4 and taking into account the local soil conditions, select a frequency for measurement or monitoring.

13.3.3 Selection of Method for Estimation

A number of methods available for carbon inventory are described in Section 13.1. The selection of the method depends on the following:

- *Size of the project* as determined by the number of sampling strata and the total number of soil samples to be analysed; remote sensing techniques could be deployed for very large projects and if the samples are few, CHN analyser could be used.
- *Accuracy and cost*, which may involve a trade-off: a CHN analyser is a very accurate but expensive; however, wet digestion is inexpensive and the most extensively adopted method.

The method to be adopted for carbon inventory has implications for costs and infrastructure requirement. Therefore, it is important to select the appropriate method for a given land-use category or project activity.

13.3.4 Selection of Sampling Technique

The sampling method adopted for above-ground biomass could be adopted for estimating SOC, assuming that biomass and SOC are linked. The sampling method and the size of the sample depend on the size of the project, number of strata and the likely spatial variation in carbon density (tonnes/hectare). Sample size could be estimated using the equation given in Chapter 10. Alternatively, the number of sampling points could be determined using the following approach:

- Adopt the "permanent plot" technique to facilitate monitoring or periodical measurements
- Adopt the sampling method used for above-ground biomass
- Select all the shrub plots used for above-ground biomass sampling, usually 8–16 plots per stratum

13.3.5 Preparation for Fieldwork

Fieldwork involves selection of soil samples for determining SOC content and core sampling for determining bulk density. The following materials are required for soil sampling.

– Soil auger	– Soil core sampler
– Bags for soil samples	– Large container for
– Field balance	– Soil sample required for determining bulk density
– Spade or shovel	
– Metallic tape	
– GPS	– Polythene cover
	– Labels/tags

13.3.6 Locating Sampling Points in the Field

Locating sampling points involves locating sampling quadrats or points using the geo-referenced grid maps.

- Obtain a map with sampling quadrats marked, if possible along with GPS points or at least with reference to some permanent and visible landmarks
- If the plots are already marked on the ground for above-ground biomass, simply locate the shrub plots
- Mark and locate the corners of shrub plots and select one of the corners for current year's sampling; the other corners and the centre point could be used for sampling in the following years
- Note the GPS reading as well as location in the format and the map for future visits
- Extract soil samples from these marked points

13.3.7 Measurement of Bulk Density Parameters

Soil bulk density is defined as the oven-dry weight of soil per unit of its bulk volume. Bulk volume comprises volume of soil solids and pore spaces, and bulk density is expressed as grams/cubic centimetres. Bulk density of soil indicates the degree of compactness and aeration, which is necessary for estimating the weight of soil per unit area, such as per hectare. Bulk density varies with soil texture – soils with fine texture tend to have lower bulk density than coarse-textured soils. – and tends to increase with depth because the lower layers are low in organic matter and microbial activity. Bulk density is considered to have relatively low spatial variability (the coefficient of variation is less than 10%) but its values are required for converting soil organic matter content to tonnes of soil organic carbon per unit area (tC/ha). Bulk density of soils is determined using the following methods (Baruah and Barthakur 1997).

(i) *Bulk density using the tube core method for undisturbed soil*

The tube core method involves sampling a soil core from a desired depth in its most natural conditions using a soil core sampler and determining the mass of solids and water content of the core; bulk density is calculated from bulk volume and weight of the dried soil.

Material and preparation The equipment needed includes a core sampler, tin sample box, balance, oven for drying samples and scale to measure dimensions of the soil core. Measurements related to bulk density could be made on the same day as that on which the soil is sampled for laboratory estimation of organic carbon.

Field and laboratory procedures The following steps could be adopted for estimating the bulk density:

Step 1: Select the same locations as those used for sampling for estimating SOC.
Step 2: Measure and record the dimensions (diameter and height) of the soil core sampler. The depth or height of the core could be 15–30 cm. Weigh the core tin box as well.
Step 3: Drive the core sampler vertically into a spot of levelled ground deep enough to fill the soil sampler tin.
Step 4: Extract the sampling core without disturbing the soil inside the core; remove any extra soil adhering to the core and protruding roots, if any.
Step 5: Weigh the sample tin along with the soil.
Step 6: Dry the soil in the tin in an oven to constant weight at 105 °C and estimate the weight of the dry soil. Ensure that the soil thus dried is not used for estimating SOC.

Data recording format for bulk density (Location, land-use category, project activity, quadrat number, sampling point number, date and GPS reading also need to be recorded)

Dimensions of the core	Length (cm): diameter (cm)
Weight of the empty tin	kg
Weight of the tin with dried soil	kg
Above-ground vegetation	Status
Location	Latitude and longitude

Calculate the bulk density (g/cc) by dividing the weight of the oven dry soil by the volume of the tin.

Bulk density (g/cc) = weight of dry soil with tin – weight of empty

(ii) *Clod method for undisturbed soil*

The clod method measures bulk density by taking an undisturbed bulk of soil clod (hence the name). Determine the volume of the clod and dry weight of the soil. Bulk density can then be calculated as the ratio of the weight to the volume. In the clod method, the volume of clod is measured by the volume of water displaced (the clod is coated with paraffin or liquid plastic before immersing it in water). Bulk density can be measured by taking an intact block of soil clod, as follows:

Step 1: Dig the soil and select a clod of soil using a pickaxe and note the depth from which the clod was collected
Step 2: Dry the clod in an oven and estimate the weight of the oven-dry clod
Step 4: Coat the clod with paraffin wax or liquid plastic
Step 5: Estimate the volume of the clod by using the water displacement method
Step 6: Estimate bulk density using the following equation

Bulk density (g/cc) = weight of the oven dried clod/ volume of the clod

(iii) *Bulk density of disturbed soil method* This method for disturbed soil involves collecting soil from known depth and filling it into a bottle or a tin and obtaining the weight and volume of the soil in the container. This method has limitations since the original degree of soil compaction cannot be simulated, leading to error. This method can be used only if other methods are not feasible.

Material A small bottle or other container (capacity about 50 ml) and analytical balance.
Procedure Weigh the empty bottle (W_1 = bottle) without the stopper. Fill the bottle with soil as explained above. Weigh the bottle again (W_2 = bottle + soil). Empty the bottle and fill it with water from a burette and record the observations:

Step 1: Weigh the empty bottle, box or tin.
Step 2: Fill this previously weighed container with soil, adding a small quantity of soil each time and tapping the container after each fill. Once filled to the brim, mark the level of soil in the container.
Step 3: Weigh the filled container.
Step 4: Empty the container and refill it with water up to the marked level. Pour the water into a measuring cylinder and note the volume (V) of water.

$$\text{Bulk density (g/cc)} = (W_2 - W_1)/V$$

13.3.8 Field Procedure for Sampling Soil for Laboratory Analysis

Collecting soil samples from the field for estimating organic matter involves the following steps:

Step 1: Locate and mark the tree sampling plots and points in the field (as described in earlier sections)
Step 2: Scrape the surface of the soil and remove litter and stones
Step 3: Using a soil augur, collect a soil sample for a depth of 0–15 cm

- Select three sampling points from each tree quadrats
- Push the augur to a depth of 15 cm
- Collect samples at this depth from all the sampling plots in the sampling stratum
- Pool the samples from the three replicates and select a subsample by successively quartering the larger sample: spread the pooled sample on a plastic sheet, divide into quarters and select any two opposite quarters; repeat the process until left with about 0.5 kg of soil
- Collect at least four to six replicates for each land-use category or project activity stratum and depth

Step 4: Repeat the procedure for 15–30 cm depth
Step 5: Transfer the soil samples immediately – within 24 h – to the laboratory to minimize loss of organic matter; if this is not possible, air-dry the samples in shade and then transfer them to the laboratory
Step 6: Dry the soil samples to constant weight in an oven at 105 °C
Step 7: Weigh the dried samples and determine the moisture content of the soil

If a core sampler or augur is not available, dig to a depth of 15 cm, collect the soil, dig deeper (up to 30 cm) and collect the second sample.

Data regarding the sample Record the following information about each sample along with date of sampling and names of field staff. A suggested format for recording details of field soil samples is shown below.

Land-use category/project activity/stratum and location	Quadrat or plot number	Sample number	Depth (cm)	GPS readings

13.3.9 Laboratory Analysis of Soil Samples

The soil samples collected have to be analysed for organic matter or carbon content in the laboratory. The bulk density of the soil also has to be determined to convert SOC content to tonnes of carbon per hectare for a given depth. Estimating SOC involves the following steps:

Step 1: Select the land-use system or project activity and the strata
Step 2: Obtain soil samples from the field along with the following information

- Location, land-use system, project activity, date of sampling
- Plot or quadrat number, replication number
- Depth of soil sample

Step 3: Select the method for estimating organic matter
Step 4: Prepare the soil samples for chemical analysis
Step 5: Follow appropriate laboratory procedure for determining the organic matter content
Step 6: Calculate the soil organic matter content using the laboratory results.

Selection of the method for estimating organic matter Various methods for estimating soil organic matter content are described in Section 13.2. The most commonly used and one with which most students and researchers are familiar is the titrimetric method, which is commonly known as Walkley and Black method or the wet digestion method.

Wet digestion method The principle and laboratory procedure is described in Section 13.2. Adopt the laboratory procedure and record the results using the following format.

Sample number	Weight of soil (*W*)	Volume of 0.5 N FAS solution used for blank titration (B ml)	Volume of 0.5 N FAS solution used for sample titration (S ml)	Volume of 1 N $K_2Cr_2O_7$ used for oxidation = 0.5 × (B–S) ml	% of organic C in soil (uncorrected)	% of organic C in soil (corrected)

13.3.10 Calculation of Soil Organic Carbon

Details of the methods for calculating SOC are presented in Chapter 17 and involves three steps. The first step involves estimation of content as a percentage. The second step requires bulk density values for the soil. The final step involves calculation of SOC as tonnes/ha using its content as percentage and bulk density.

13.3.11 Long-Term Monitoring of Soil Organic Carbon

Soil organic carbon accumulation or loss occurs over long periods, extending over decades. Therefore, it is necessary to monitor soil organic carbon stocks periodically. The frequency of monitoring is normally once in 3–5 years (Chapter 4). Adopt the permanent plots used for monitoring above-ground biomass for periodic sampling and monitor the soil organic carbon using the procedure described in Section 13.3.

13.4 Conclusions

Soil organic carbon is a key carbon pool for many land-use categories and project activities, the stock of which depends on the type and status of vegetation. Mineral soils dominate the land-use systems and only organic carbon is impacted by land-use changes and management systems. Thus the focus is on SOC and mineral soils. Multiple methods are available for estimating SOC, of which the wet digestion method is the most common whereas CHN analyser, although the most reliable, is expensive. This chapter presented field methods of soil sampling and laboratory methods of estimating organic carbon content. Soil carbon needs to be estimated for most land-use categories and project types, except where soil is not disturbed and only management practice is changed, for example, improved forest management, with forest land remaining forest land. Soil organic carbon is impacted by changes in land-use that disturb the topsoil, such as forest land or grassland converted to crop land but is stable otherwise. The aim of most carbon mitigation, roundwood production and land reclamation projects is to increase the carbon stocks in soil, which makes estimation of SOC a necessity for most projects. Soil organic carbon is also one of the key indicators of soil fertility and thus of interest to most mangers of land-based projects. The methods described in this chapter could be adopted for estimating carbon stocks for national greenhouse gas inventories.

Chapter 14
Remote Sensing and GIS Techniques for Terrestrial Carbon Inventory

Remote sensing is a technique that holds great potential for long-term monitoring of changes in area and carbon stocks. This chapter discusses the application of different techniques for different project types in terms of feasibility and reliability; highlights uncertainties, cost and required technical capacity; describes the application of geographical information systems (GIS) methods for carbon inventory for different projects; and also assesses the role of remote sensing and GIS techniques for long-term carbon inventory.

Data from remote sensing are those acquired by sensors, which may be in the form of optical devices, radar or lidar on-board satellites or cameras equipped with optical or infrared films installed in aircraft. The data can be thought of as an image representing the ground. These data can be used to provide estimates of land cover and area although, being essentially interpretation of images, it is usually necessary to validate the data against ground truth to know how accurate the interpretations are (IPCC 2006). Remote sensing is a powerful tool in that it covers large areas and enables inventories to be made at a low cost per unit area. Analysis of satellite imagery is the most practical approach to monitor vegetation cover over large areas periodically as a routine (DeFries et al. 2005). If sample plots with adequate ground-based measurements are insufficient to support a proper inventory over a large area, it is necessary to use auxiliary variables correlated with land-use variables. One such variable can be obtained by remote sensing or GIS (Lappi and Kangas 2006). Another benefit is the savings since field measurements are one of the most expensive components of sampling-based land-use inventories for large areas (Tomppo 2006).

Data from remote sensing can be classified either by visual analysis of the imagery or by digital, computer-based methods. The strength of remote sensing is its ability to provide spatially explicit information repetitively. Archives of remote sensing data also span several decades and can therefore be used to reconstruct land cover and land use as a time series. Remote sensing is particularly useful in obtaining estimates of area under different types of land cover and land-use categories as described in Chapter 8. Furthermore, remote sensing can assist in identifying and defining relatively homogeneous areas that can guide in sampling (see Chapter 10 for information on sampling design and size of samples).

The challenges for remote sensing are in interpretation, which is the process whereby images or data are translated into meaningful information on, for example, land cover and land use. A common obstacle is the presence of clouds, aerosols and haze, which limit the availability of data to varying extent, depending on the sensor used. A radar, which actively sends out a signal that bounces back from the surface, is not limited by these factors whereas passive sensors that rely on the actual reflectance from the surface are hampered by such obstructions. Another obstacle is the difficulty in distinguishing between different land-use or cover types that give very similar signals. Another concern when comparing data over long period of time is that remote sensing systems may have changed over time in terms sensors, bandwidth or maintenance.

14.1 Implications for Carbon Inventory

A carbon inventory requires estimates of biomass stocks, and remote sensing helps in generating information or data required for such estimates or for validating the estimates made by other means. Estimates of biomass stock are based on features of vegetation such as coverage and canopy.

One of the crucial issues in using data from remote sensing is the accuracy of estimates. The concern also extends to assessment of carbon inventories. Remote sensing, attempts to correlate a spectral signature with a specific land use, and it is necessary to define how closely the interpretation matches reality (UNFCCC 2006). Accuracies of 80–95% are achievable with high-resolution imagery (where every pixel covers only a small area) in discriminating between forest areas and non-forest areas. However, to detect biomass and hence the carbon content, such high accuracy is much harder to attain – the most reliable estimates of carbon stocks are those based on field measurements.

Carbon is the main component of biomass vegetation and is invisible. Therefore, it is necessary to focus on features of vegetation to estimate carbon stocks. These features can be age of the vegetation (see, e.g. Zheng et al. 2004), tree diameter (see, e.g. Drake et al. 2003), intensity of chlorophyll activity or biomass density (see, e.g. Tan et al. 2007).

14.2 Data from Remote Sensing

Remote sensing is the process of obtaining information about an object, area, or phenomenon through the analysis of data acquired by instruments not in direct contact with the object being investigated. Reading, for example, is a remote sensing process. Eyes act as sensors that respond to the light reflected from a page in a book. The data that the eyes acquire are impulses corresponding to the amount and pattern of light reflected from the dark and light areas on the page. These data are analysed or interpreted by the brain to enable a person to explain the dark areas on the page as a collection of letters forming a word (Lillesand and Kiefer 1994).

At present there are about 800 satellites operating with the aim of collecting information on a variety of environmental topics, such as the atmosphere, snow, oceans and vegetation. The satellites move in different types of trajectories, orbits or paths, such as geostationary or polar satellites. Most of the sensors placed on satellites for earth observations are on polar-orbiting satellites at a height of 450–900 km, usually synchronized with the Sun for visibility since the optical sensors need light to collect information.

A remotely sensed image is a pixel-by-pixel measurement of reflected or emitted energy from the Earth's surface (Brown 1997). A few of the most commonly used types of remote sensing data are aerial photography, satellite imagery using visible and/or near infrared bands, satellite or airborne radar imagery and lidar. Combinations of different types of remote sensing data may very well be used for assessing different land-use systems or areas and thus for estimating carbon stocks. These combinations can include interpretations of two data sets to increase accuracy and the use of two or more bands to produce indices. One set of such useful indices comprises vegetation indices using both visual and infrared bands. A system based on remote sensing to track land-use conversions can include many combinations of sensors and data types at a variety of resolutions.

There are several important criteria for selecting remote sensing data and products for terrestrial carbon inventory (IPCC 2006):

1. *Adequate land-use system stratification scheme* Stratification of the project area has to be robust and clear to be able to distinguish between them. The stratification should be of adequate spatial resolution to enable use of remote sensing.
2. *Appropriate spatial resolution* If broad categories or distinct land-use differences are sought, such as forested and non-forested land, low-resolution remote sensing might be adequate, compared to a detailed categorization of different agricultural land that requires high resolution.
3. *Appropriate temporal resolution* Estimating land use changes in boreal forest systems might require data that span over decades, whereas for estimating changes in grassland, data for even a single year may be sufficient. Seasonality of the vegetation is an important factor since peak vegetation period is usually the best time for inventory of terrestrial carbon.
4. *Availability of historical assessment* Often the limitation of conducting a remote sensing survey is the availability of historical data. In that sense the future is promising, since more, readily available, sensors and products are being developed.
5. *Transparent and consistent methods applied in data acquisition and processing* Since carbon inventories are performed frequently and require monitoring over time, the methods that are used have to be repeatable.
6. *Consistency in data and availability over time* The products used should be consistent over time for the same reason as stated in point five above.

Different remote sensing sensors are receptive to energy from diverse parts of the electromagnetic spectrum, such as visible, near-infrared, infrared or thermal.

Sensors collect parts of the wavelength into different bands or information sets, which means that over a particular pixel, several bands are produced for the same area, which can be used to construct indices for numerous features such as vegetation types. The process is described later in this chapter. Different features such as forest, bedrock, soil or cropland have different reflectance and it is the spectral differences between these features that enable the user to classify the image into different land-use types (Brown 1997). More information on the elements of photographic systems related to remote sensing can be found in textbooks such as that by Lillesand et al. (2004).

Classification of land use with remotely sensed data can be achieved visually or digitally; the latter essentially means computer-based analysis. Each approach presents advantages and disadvantages. Visual analysis allows for human inference through the evaluation of overall characteristics of the image. Usually, this is done by analysing the contextual aspects of the image. Digital classification, using computer hardware and software, allows for several manipulations to be performed with the data, such as merging of different spectral data (Fuentes et al. 2006) and adding information from ancillary data from object-oriented methods (Bock et al. 2005), which can help to improve modelling of biophysical ground data, such as tree diameter, height, basal area, biomass, time of flowering and harvest or disturbances such as drought, diseases or fire. Digital analysis allows immediate computation of areas associated with different land-use categories and has developed rapidly over the past decades; given the concurrent development related to computers, the necessary hardware, software and also the satellite data are now readily available at low cost.

Aerial photography Analysis of aerial photographs can reveal differences in land-use or land-cover system such as agriculture, grassland, forest tree species and forest structures from which the relative distribution and tree health may be judged. In agriculture, similar analyses can show crop species, crop stress or tree cover in agroforestry systems (Fig. 14.1). The smallest spatial unit that can be seen depends on the type of aerial photos used, but for standard products it is often as small as one metre (IPCC 2006).

Optical satellite images Complete national or regional land-use and land-cover analyses may be facilitated by satellite images. This section describes passive satellite data in the visible and near-infrared spectra. Passive sensors rely on reflectance of solar energy from the surface back to the sensor or detector. This energy is captured in the visible, near- and middle-infrared portion of the electromagnetic spectrum (~0.4–2.5 μm). Digital multispectral remote sensing data record spectral information in a number of wavelengths referred to as bands. Information up to 10 bands per pixel or unit of land can be recorded. Hyperspectral data, consisting of 100–200 bands of information, are available but require special processing methods (see, e.g. Tamás and Lénárt 2006). Green vegetation exhibits a unique signature characterized by strong reflectance in the green and infrared portions of the electromagnetic spectrum and strong absorbance in the red and some mid-infrared regions (see Fig. 14.2). Variations in internal cell structures of leaves, absorption levels of chlorophyll and variations in leaf water content make it

Fig. 14.1 An aerial photograph with 4 m resolution over Julita, in mideastern Sweden (~2 × 2.5 km). The photo is an orthophoto, meaning that it has been geometrically corrected. Agriculture, deciduous forest and shelterbelt trees can be detected as well as individual houses and gardens. The area in the lower left corner is part of Lake Öljaren. (Courtesy of Lantmäteriet Gävle 2007. Medgivande I 2007/437)

possible to distinguish between different types of vegetation (Patenaude et al. 2005). Various indices, such as normalized difference vegetation index (NDVI), have been designed to optimize these spectral signatures of vegetation.

Time series can be obtained for any area of interest since the satellite passes over it continuously and regularly. Reliable optical data going back to the early 1990s can be accessed and used for interpretations with good confidence for assessing changes in land use (DeFries et al. 2006). Because satellites circle the globe in different orbits at different heights and speeds, the interval between consecutive passes over one particular geographical area differs between satellites. The images usually generate a detailed mosaic of distinct categories, but matching them to proper land cover and land-use categories commonly requires ground reference data from maps, field surveys or other available information.

The smallest unit to be identified depends on the spatial resolution of the sensor and the scale of work. The most common sensor systems have a spatial resolution of 20–30 m. At a spatial resolution of 30 m, units as small as 1 ha can be identified. Data from higher resolution satellites are also available (IPCC 2006). Several satellite products are presented in Table 14.1 with a range of resolution from 0.6 to 1,100 m, from fine and high to coarse and low resolution. Many of the satellite products can be accessed through official web sites.

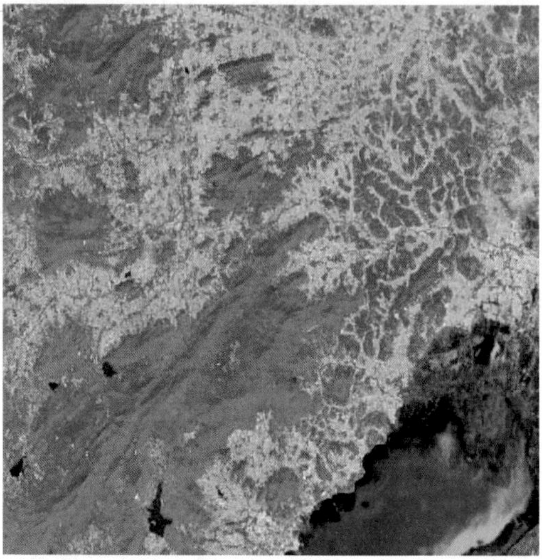

Fig. 14.2 An optical 20 m resolution SPOT multispectral image (228/309) over Western Orissa, India, on 29 December 1994 (~60 × 60 km). The image consists of information from all bands, from visible green and red to near-infrared, also called a false colour composite image, which makes green vegetation appear red, a feature that is important for carbon monitoring

Table 14.1 Examples of passive satellite images. (From UNFCCC, 2006.)

Satellite (sensor)	Resolution (m)	Time coverage	Cost	Source
NOAA (AVHRR)	1,100–8,000	1978–	Free	http://edc.usgs.gov
EnviSAT (MERIS)	300–1,200	2002–	$525/scene	http://envisat.esa.int/
Terra (MODIS)	250	2000–	Free	http://edc.usgs.gov http://glcf.umiacs.umd.edu/data/gimms/
Landsat (MSS)	60	1972–1992	Free to $375/scene	http://www.spaceimaging.com http://edu.usgs.gov http://glcf.umiacs.umd.edu/data/gimms/
Landsat (TM)	25	1982–	Free to $625/scence	http://www.spaceimaging.com http://edu.usgs.gov http://glcf.umiacs.umd.edu/data/gimms/
Landsat (ETM+)	15	1999–	Free to $800/scene	http://www.spaceimaging.com http://edu.usgs.gov http://glcf.umiacs.umd.edu/data/gimms/
SPOT (VGT)	20 (10)* 2.5	1986–	$1,200–10,125/scene	http://www.spotimage.fr/home http://www.spot.com
Terra (ASTER)	15	1999–	$145–580/scene	http://edc.usgs.gov
IKONOS	4 (1)*	2000–	$16–56/km²	http://www.spaceimaging.com http://glcf.umiacs.umd.edu/data/gimms/
Quickbird	2.4 (0.6)*	2001–	$5,000–11 500/scene $16–45/km²	http://www.digitalglobe.com http://glcf.umiacs.umd.edu/data/gimms/

*Available in panchromatic, which means it is in black and white

The relationships established between vegetation biomass and hence carbon stocks and the data from optical sensors with low resolution are generally weak (Rosenqvist et al. 2003). Many of the early or pre-2000 satellite products are low to medium in resolution, a fact to be taken into account when seeking historical information from optical satellite imagers on carbon stocks.

Radar images Unlike optical satellites, which rely on solar illumination, radar, an acronym for radio detection and ranging, are active microwave sensors that emit energy to survey the Earth (Lillesand et al. 2004). A major advantage of a radar system is that it can penetrate clouds, aerosols and water vapour (Fig. 14.3). Radar also acquires data during the night, whereas passive and optical products cannot do so. Therefore, the system is called an active satellite system since it sends out a signal that hits the surface and returns to the sensor. This makes radar perhaps the only reliable source of remote sensing data in many areas of the world with frequent cloud cover, such as the tropics.

The most common type of radar data is the so-called synthetic aperture radar (SAR) sensor system that operates at microwave frequencies, such as RadarSAT. By using different wavelengths and different polarizations, SAR systems may be able to distinguish between land-use systems, for example, forest and non-forest, or the biomass content of vegetation. The radar system transmits either horizontally (H) or vertically (V) polarized electromagnetic (EM) energy and then receives either of these polarizations. The frequency transmit/receive configuration of radar data is typically stated by a three-letter code: the first letter designates the band of the radar and the last two letters state the polarization configuration. Four combinations are in use, namely HH, HV, VH and VV. L-VH radar, for example, means an L-band system that transmits vertically and receives horizontally in terms of polarized EM energy (Kasischke et al. 1997). At present radar has some limitations when biomass is high due to signal saturation, that is the signal does not change beyond a certain chlorophyll level.

Fig. 14.3 Two radar images, namely synthetic aperture radar (SAR) and phased array type L-band synthetic aperture radar (PALSAR) showing deforestation in Amazon from 1996 to 2006. Grey indicates forested area and black represents deforested area. (Courtesy of Japan Aerospace Exploration Agency's Advanced Land Observing Satellite (JAXA/ALOS))

In recent years, SAR backscatter has been increasingly investigated for use in inventorying forests because radar wavelengths penetrate the vegetation and, in that sense, provide direct feedback about vegetation structure and biomass (Patenaude et al. 2005). The most commonly used systems are the C-band (wavelength ~5 cm), L-band (~24 cm) and P-band (~70 cm) (Kasischke et al. 1997; Igarashi et al. 2003; Lucas et al. 2006). The short wavelength, C-band, is sensitive to small components of the canopy such as leaves and twigs whereas the other two bands penetrate deeper and are predominantly sensitive to larger branches and trunks.

One limitation of radar SAR systems is their sensitivity to surface topography, which limits their general application to flat or gently undulating terrain (Rosenqvist et al. 2003). The main external factor controlling the sensitivity of the SAR signal to biomass is the structural properties of the forest or vegetation and its underlying surface. Given equal biomass, the radar backscatter response will be significantly different for a forest composed of sparsely distributed large trees and one composed of dense stands of young, small trees (Patenaude et al. 2005). Examples of available data collected using active radar remote are listed in Table 14.2.

In response to the demand for technical and scientific input to the work on climate change, carbon inventory and the Kyoto Protocol, the Kyoto & Carbon initiative was initiated by Japan Aerospace Exploration Agency (JAXA) as part of the Advanced Land Observation Satellite (ALOS) in 2000. One of the focal points was to utilize ALOS phased array type L-band synthetic aperture radar (PALSAR) to support the type of information needs on a regional scale associated with carbon estimations (ALOS 2006). As of now (early 2007), the products generated from this effort are made available to general users about 6 months after initial distribution to the Kyoto & Carbon Science team.

Lidar Light detection and ranging, or lidar, uses the same principles as radar. The lidar instrument transmits light to the target; the transmitted light interacts with the

Table 14.2 Examples of satellite images taken with active radar systems. (From ALOS 2006; Rosenqvist et al. 2003; Lillesand et al. 2004; UNFCCC 2006.)

Satellite (sensor)	Resolution (m)	Time coverage	Cost	Source
ENVISAT (SAR, ASAR MERIS)	25–150	2002–	$150–1,000/scene	http://envisat.esa.int/
ERS-1 (SAR)	25–150	1991–2000	$150–700/scene	http://www.esa.int
ERS-2 (SAR)	30	1995–	$150–700/scene	http://www.esa.int
JERS-1 (SAR)	18–100	1992–1998	$100–1,000/scene	http://www.Eorc.nasda.go.jp/JERS-1
ALOS (PALSAR)	9–157	2006–	Free to $250/scene	http://earth.esa.int/dataproducts/
RadarSAT 1 (SAR, ASAR)	8–100	1995–	Free	http://www.rsi.ca
RadarSAT 2 (SAR, ASAR)	3–100	2004–		http://www.rsi.ca

target and is changed by it. Some of this light is scattered and reflected back to the instrument, which analyses the light. The change in properties of the light enables some properties of the target to be determined. The time taken by the light to travel to the target and back to the lidar is used to determine the distance to the target (IPCC 2006). Lidar resolution or footprint size may vary from 0.25 to 25 m (Drake et al. 2003; Rosenqvist et al. 2003), which makes lidar more detailed than the other technologies described so far. Products of lidar come in a wide range, although most of them are commercial, and therefore not as readily available as those from optical and radar data. Lidar is at present not available from satellite platforms, which limits its use even more (Patenaude et al. 2005). Irrespective of the type of lidar instrument used, the general approach has been to use some physical attribute a forest canopy (Næsset 2002), such as canopy height (Kimes et al. 2006), tree height and stem volume (Holmgren et al. 2003) and canopy elements in three dimensions (Lovell et al. 2003) to estimate biomass, particularly above-ground biomass (Lim and Treitz 2004). Lidar is the youngest technology of gathering data among the remote sensing ensemble, which can be used in characterizing vegetation. Lidar products have demonstrated a strong relationship between physical features such as tree height, stem volume, biomass and canopy closure (Drake et al. 2003; Rosenqvist et al. 2003).

The limitation of lidar is that it is a fairly sophisticated technology, which requires trained experts for analysis not available everywhere. Also, at present, it is too expensive to be applied over large areas (Skutsch et al. 2007). Another limitation is lidar's inability to distinguish between species of trees in a forest. As wood density and hence biomass varies between tree species of similar height and age, estimates of biomass using lidar alone may be less accurate (Rosenqvist et al. 2003).

Laser The concept of laser is yet another active system, where a spot on the Earth's surface is illuminated by a laser beam and the distance to the spot determined. Laser as a remote sensing technique is considered a promising tool for monitoring changes in vegetation changes such as avoided deforestation (Joanneum et al. 2006). The geosceince laser altimeter system (GLAS) sensor on-board the Ice, Cloud and Land Elevation satellite – Icesat – was primarily aimed at monitoring mass balance of the polar ice sheet but has also proved useful in assessing vegetation. GLAS produces a series of spots, ~70 m in diameter, which are collected in a telescope 1m in diameter, which gives high-resolution data for analyses. The satellite was launched in 2002 but started operating in 2004.

14.2.1 Remote Sensing and Ground Reference Data

In using data from remote sensing for inventories and particularly in studying issues related to land use, it is a good practice to complement the data with ground reference data: data from remote sensing need to be validated against empirical data that can be collected independently or obtained from forest and agricultural inventories. Data from remote sensing, and analyses based on such data, are always interpretations, the accuracy of which can be increased when validated with field information. This cross-check should be done regardless of resolution and source.

Land-use systems that are rapidly changing over the estimation period or that have features known to be easily misclassified should be more intensively ground-truthed than other areas. This can only be done by using ground reference data, preferably from independent actual ground surveys. High-resolution photographs may also be useful.

14.2.2 Calibrating Remote Sensing Data

The data acquired from remote sensing need to be calibrated. The broad categories of calibration are radiometric, atmospheric and geometric. Calibration applies to data that have not been pre-processed. However, many products available today are pre-processed and therefore can be adopted easily; if not, the information required for calibration is usually given by the distributor.

1. Radiometric calibration relates to the conversion of raw digital data into spectral radiance (Lillesand and Kiefer 1994)
2. Atmospheric calibration relates to "noise" in the data, which is a result of such components of the atmosphere as aerosols, water vapour and smog (Brown 1997)
3. Geometric calibration relates to the way a curved area is to be presented in a flat format, taking into account possible distortions in angles due to the location of sensors in relation to what has been observed.

14.3 Methods to Estimate Biomass

The carbon pool that is most accurately estimated through remote sensing is above-ground biomass, which is described in this chapter. It is possible to estimate below-ground biomass from the data on above-ground biomass derived from remote sensing (see Chapters 4 and 11).

Depending on the type of land-use system being monitored for estimating carbon stock, different features of change in vegetation over time have to be considered (Rosenqvist et al. 2003); these include different types of harvesting and seasonal changes. For repeated carbon inventories, it is also important to select the same period of the year for estimation, usually the peak vegetation period.

Various methods are available to interpret and analyse satellite data for measuring changes in land cover and hence biomass. However, no consistent method or technique is available for estimating carbon through remote sensing (Rosenqvist et al. 2003). The methods range from visual interpretation of photographs to sophisticated digital analyses and from "wall-to-wall" mapping (covering a contiguous stretch of land such as a province, country or continent) to analysis of hot spots and statistical sampling. A variety of methods can be applied depending on technical capabilities and characteristics of the land-use pattern. There are several conventional methods to estimate above-ground biomass through remote sensing (Labrecque et al. 2004), such as radiometric relationships between satellite reflectance or spectral indices and biomass values measured on forest inventory plots, nearest-neighbour approaches, unsupervised classification of land cover and forest structures characteristics and forest sample plot databases. Two methods including the steps in estimating carbon stocks by using remote sensing are presented in the following section.

It is not possible to directly measure the total stocks in above-ground biomass or changes in the stocks through remote sensing. In quantifying biomass and thus carbon stocks, remote sensing data are used in combination with empirical data, either directly using allometric relationships or indirectly based on features such as canopy cover. Reflectance in different parts of the spectrum, either alone or in combination such as indices or principal components, with strong empirical relationship, can also be used for estimating biomass. Indirect estimates using empirical relationships including canopy cover, indices from several bands, photosynthetically active radiation (PAR) or net primary production (NPP) that combine environmental data with remotely sensed data are usually necessary. Additional methods include quantification of productivity through the use of light use efficiency (e.g. Brogaard et al. 2005).

In combination with empirical biomass measurements, remote sensing data can be used to extrapolate over larger areas and also over longer time frames. Usually, this is done by developing a regression model between the empirical data and the remote sensing data (e.g. Dong et al. 2003).

14.3.1 Estimating Biomass Based on Remote Sensing Vegetation Index

Indices based on remote sensing data are a set of useful techniques to be used for estimating biomass. Some examples of indices are as follows: NDVI, or normalized difference vegetation index (Dong et al. 2003; Zheng et al. 2004; Fuentes et al. 2006; Tan et al. 2007); EVI, or enhanced vegetation index (Huete et al. 2002; Nagler et al. 2005; Ostwald and Chen 2006); LAI, or leaf area index (Fassnacht et al. 1997); PAR, or photosynthetically active radiation (Wylie et al. 2007) and FPC, or foliage projected cover and CPC, or canopy projected cover (Rosenqvist et al. 2003). Lu et al. (2002) gives examples of several more vegetation indices used in the Amazon. These indices can be used in combination with other techniques such as field measurements or other environmental data to get carbon content of a land-use system.

As the first illustration of steps involved in estimating biomass, NDVI is used here with forest vegetation. NDVI uses the ratio of the near-infrared and the red spectra (NIR – red/NIR + red) and can be used as a proxy for green leaf area (Myneni et al. 1998) since it signals photosynthesis and is often used as an index to reveal seasonal and/or inter-annual change in vegetation cover. NDVI has been widely used for biomass studies because of its availability and long history of vegetation measurements (Todd et al. 1998; Dong et al. 2003; Seaquist et al. 2003; Zheng et al. 2004; Fuentes et al. 2006; Myeong et al. 2006; Tan et al. 2007; Wylie et al. 2007).

Step 1: Collect inventory data for biomass of the forest

- Depending on spatial scale and accuracy needed, useful data can be collected from a plot size of 50 × 50m (Lu et al. 2002; Lucas et al. 2006) to that from statistics for an entire province (Dong et al. 2003)
- Geoposition the data to make it correspond to the remote sensing data
- Convert the data into carbon (see Chapter 10 for methods)
- If there is large amount of data, split them into a model-development set and a validation set (Labrecque et al. 2004)

Step 2: Collect NDVI data for the area

- NDVI products do come ready-made from several databases but can also be obtained from data having bands covering the near-infrared and red spectra
- Resolution should, as far as possible, correspond or be smaller than the spatial coverage of the inventory data
- Techniques for stepwise nesting using different resolutions of remote sensing data are described by Muukkonen and Heiskanen (2006)
- One useful method is to collect NDVI data for several years preceding the date of biomass inventory covering the growing season of the forest (Dong et al. 2003). Dong et al. (2003) suggest cumulating the NDVI for the growing season

Step 3: Make sure the remote sensing data are calibrated to reduce satellite or technical noise or atmospheric issues (see Section 14.2.2)

- Most common are cloud effects (Lillesand et al. 2004)
- Data distributors usually have information on how this can be done

Step 4: Make sure the NDVI data are georeferenced so the inventory plots can be identified in the corresponding pixels

Step 5: Produce a NDVI model

- Parameters to be considered in the NDVI model are composition of species (Labrecque et al. 2004), age of stand (Zheng et al. 2004), unsupervised classification of land-use classes (Labrecque et al. 2004), latitude if the area is covering a large area (Dong et al. 2003), vegetation texture (Lu et al. 2002) and logarithmic forest biomass (Tan et al. 2007)

Step 6: Find the relationship between carbon data and NDVI

- Correlate the two data sets to develop a statistical relationship. Examples of tests are Pearson's correlation coefficient and stepwise regression analysis (Lu et al. 2002).
- If using two sets of data, one for developing the model and one for validation, a simplified error matrix can be used (see Table 14.3).

Table 14.3 Error matrix table and evaluation of accuracy based on modelled and evaluated sets

Measured	Error matrix				
Modelled	Low C	Medium C	High C	Total	Producer's accuracy (%)
Low C	180	50	38	268	180/268 = 67
Medium C	49	210	31	290	210/290 = 72
High C	40	30	200	270	200/270 = 74
Sum	269	290	269	828	Overall accuracy 590/828 = 71

Step 7: Evaluate the relationship

- Coefficient of determination (R^2) describes the percent of variation explained in the regression model, with stronger correlation between variables with values close to 1. See Section 14.4 for accuracy of different remote sensing applications
- If using a model development set and an evaluation set, percent accuracy is given as stated in the error matrix (Table 14.3).

NDVI is used here to demonstrate the procedure, but other indices or single spectral bands can be used and tested depending on their availability and requirements of the project. These steps are fundamental, and irrespective of the type of sensor or type of remotely sensed data being used.

14.3.2 Estimating Biomass Based on Remote Sensing-Derived Land-Use Change Classes and GIS

A combination of visual interpretation and digital classification can many times be a good method to generate information for an area (Bickel et al. 2006). Remote sensing can be used to detect locations of change when using data from two different periods. The method contains two categories, which have been used in the 2006 IPCC Guidelines for National Greenhouse Gas Inventories (Bickel et al. 2006).

Post-classification change detection approach The approach of detecting post-classification change refers to techniques where two or more predefined land-use classifications exist for different points in time and where the changes are detected, usually by subtraction of the data sets. The techniques are straightforward but are also sensitive to inconsistencies in interpretation and classification of land-use categories.

Pre-classification change detection approach The approach of detecting pre-classification change refers to more sophisticated and biophysical approaches to detecting change. Difference between data from spectral responses from two or more periods are compared statistically and used to obtain information on land-use change. This approach is less sensitive to inconsistencies in interpretation and can detect much more subtle changes than is possible with the post-classification approaches, but is less straightforward and requires access to the original remotely sensed data. Simple visual interpretation can be used to assist these two approaches.

Areas of change are highlighted through display of different band combinations, band differences or derived indices, such as NDVI.

The following steps illustrate the approach to estimate changes in biomass stocks using land-use changes detected through remote sensing. The post-classification change detection is used here to estimate biomass and hence carbon stock:

Step 1: Collect land-use information for the area

- The information can be in the form of maps or data that can be used as a check against satellite-derived classification

Step 2: Collect remote sensing data for the area

- Several different remote sensing products can be used; Landsat is probably the most common because of its temporal coverage and availability (see Table 14.1 for web site information)
- Several bands can be used, where visible and near-infrared spectra have proven suitable for identifying vegetation

Step 3: Make sure the remote sensing data are calibrated to reduce satellite or technical noise or atmospheric issues

- Data distributors usually have information on how this can be achieved

Step 4: Make sure the data are georeferenced so the inventory plots can be identified in corresponding pixels

Step 5: Produce land-use classes using the remote sensing data

- Perform unsupervised classification using image processing programmes or GIS

Step 6: Evaluate the classification with the help of land-use information

- This can be done by using the error matrix (see Table 14.3)

Step 7: If the relationship is low, try other types of land-use classification

- The accuracy should correspond to the certainty needed for the project

Step 8: Calculate biomass based on the inventory data and areas under different land-use classes

Saturation effect Saturation effect with remote sensing occurs when vegetation is dense and the added biomass, chlorophyll or leaf layers are not shown in the obtained data. This makes the assessment of heavy vegetation more unreliable. In adopting NDVI, which ranges from −1 to 1, it is suggested that the added chlorophyll activity after 0.7 is uncertain. When measuring biomass, it is suggested that saturation occurs in optical remote sensing at about 15 kg/m^2 (Steininger 2000). For several space-borne SAR radar products, the saturation is slightly extended to ~20 kg/m^2. This limits the use of data for routinely quantifying biomass, particularly as the majority of forest and woodlands globally supports an above-ground biomass of more than 100 t/ha or 10 kg/m^2 (Rosenqvist et al. 2003).

14.4 Uncertainty and Accuracy

Whenever a map of land use is used, it is necessary to know the reliability of the information presented in the map. When results are generated from classification of remote sensing data, it should be recognized that the reliability of the map is likely to vary for different land-use categories: some categories may be uniquely distinguished while some may be confounded with others. For example, coniferous forest is often more accurately classified than deciduous forest because reflectance characteristics of a coniferous forest are more distinct. Similarly, it is often difficult through remote sensing to ascertain changes, such as a change from intensive tillage to reduced tillage, in land management practices, in a specific land area (IPCC 2006). The level of detail can increase by subdividing, for example, dividing cropland by crop or a forest into different basal area classes, depending on what is being monitored and to what degree of accuracy. To evaluate the accuracy of mapping the following steps for basic interpretation could be adopted:

Step 1: Select a number of sample plots from each land-use type shown on the map based on the interpretation. Record the coordinates for the plots

Step 2: Collect actual real-world data on land use from some type of ground-truthing data

Step 3: Create a matrix (see Table 14.3) of interpreted/modelled and actually measured land-use types

Step 4: Calculate percentage of accuracy from the matrix.

Age of vegetation developed as ascertained from field measurements in combination with information obtained from near-infrared remote sensing from a Landsat ETM + image could give estimates of above-ground biomass of hardwood forest stands in northern US with a coefficient of determination (r^2) of 0.95 (Zheng et al. 2004).

Uncertainty in relation to heterogeneity of land-use types The accuracy of interpretation is related to the homogeneity of the surface of vegetation being investigated. Remote sensing data from a large area of boreal forest are less variable than those from a dry tropical forest with hundreds of different tree species. Similarly, estimates of production of wheat from large-scale cultivation are likely to be more reliable than those of production of rice from rice paddies scattered among blocks of agroforestry. Differences in soil moisture, topography and patches of atmospheric disturbances such as clouds of haze can also lower the certainty of interpretation from remote sensing data.

Uncertainty in relation to topography Topographic features in an image prevent light or radar energy from being reflected back to the remote sensing detector. In an image of mountainous areas, there are large dark areas such as valleys and shadowed ridges. When using data with small pixels size, the impact of shadows is larger than that with large-scale or coarse images. For example, 50 m pixels reported a 14% error rate, which dropped to 3% in 1.1 km pixels (Brown 1997).

14.5 Feasibility of Remote Sensing for Different Project Types

Methods to distinguish between forest and other land-cover types using remote sensing data are fairly accurate when the contrast between them is high. An accuracy of 80–95% is expected with high-resolution images. The problem arises when other land-cover types also have green vegetation, perhaps even trees. Forest parameters to be determined from the image, such as canopy cover and extent of degeneration that are supported without ground-truth data are not readily available. The extent and scale of ground-truthing is determined by the resources available.

Afforestation and reforestation Application of remote sensing for afforestation and reforestation projects is feasible since the main feature of project implementation is conversion of different land-use systems to forests and plantations. Distinction between clearly distinct land-use systems, such as forest and non-forest is more reliable than interpreting the difference between young and mature forest stands or that of partly degraded forest and non-disturbed forest. CDM sink projects under the Kyoto Protocol are required to prove that the land was not forested in the past. Remote sensing can play an important role in this process, since data are available at least as far back as the early 1980s.

Avoided deforestation Annual emissions from land-use change, mainly through deforestation and degradation in tropical developing countries, account for ~20–25% of the total anthropogenic emissions of greenhouse gases. The uncertainty of the estimate is varied, given the range. The factors contributing to uncertainty are lack of resources, lack of standard methods and lack of data and capacity at the national level. Standardization is required for using remote sensing data, tools and analytical methods that suit the variety of national conditions as well as meet acceptable levels of accuracy (UNFCCC 2006).

The role of remote sensing will be central for projects aimed at avoiding deforestation or reducing emission from forest degradation in the tropics, since the area to be covered is likely to be large and inaccessible in places. Apart from conversion of forest to non-forest uses, the process of degradation, thinning or regeneration is also important since degradation of forest decreases carbon content. The idea of reducing the emissions from these areas automatically calls for a baseline and a base period. A historical baseline could be constructed on the basis of area under forest cover and extrapolated to the future based on remote sensing data (Skutsch et al. 2007).

Deforestation and forest degradation need to be differentiated since both contribute to loss of carbon stocks. Remote sensing technique has the potential to help in distinguishing the two processes at a much lower cost than other methods.

In addition to monitoring land use and changes in land cover, remote sensing techniques are particularly useful in integrating other factors such as population density, markets and ownership (Skutsch et al. 2007) that drive changes in land use and land cover. By adopting a GIS approach, issues related to remote sensing can be incorporated with the other spatially determined features to be used in analysis and monitoring of avoided deforestation (Castillo-Santiago et al. 2006). Several models have been developed to deal with avoided deforestation and related management issues.

14.6 The Role of GIS

Visual interpretation of images is often used in identifying sampling sites for ground inventories. The method is simple and reliable. However, it is labour intensive and therefore restricted to limited areas and may be affected by subjective interpretations by different interpreters. There has been a revolution in the way information about environment is acquired, processed and stored, which has been attributed to advances in computers for data collection and manipulation. Geographic information systems, or GIS, have played a key role in the development (Rosenqvist et al. 2003). The power of GIS comes from a database management system that is designed to store and manipulate data (Lillesand et al. 2004).

Application of remote sensing requires integration of the extensive remote sensing data with ground-truth measurements or data to characterize areas associated with multiple features. This is generally achieved most cost effectively using a GIS (IPCC 2006). Apart from its application for remote sensing data, GIS also offers the possibility of integrating for further analysis other types of information including data on soil types, population of a certain area and infrastructure or management practices, as presented in Fig. 14.4 (Ostwald 2002).

14.7 The Future for Remote Sensing and GIS

Technological advances in remote sensing techniques will enhance their use in many different land-use monitoring applications. The data are getting more reliable, available and affordable. Avoided deforestation is one area where remote sensing is being targeted as the key approach to holistic and acceptable methods of monitoring changes. These changes extend beyond the straightforward "forest to deforestation" process to also include degradation of forest resources (DeFries et al. 2006). Modelling could be employed to increase accuracy and reduce the cost of estimating carbon stocks. Over the last two decades, many biophysical models of forest growth dynamics have been developed, many of them with the specific objective of using data from satellite images as input to drive the models (Porté and Bartelink 2002; Skutsch et al. 2007).

Recent developments in remote sensing technology have advanced its application in estimating carbon stocks in land-use systems. Here, radar and laser systems are the most promising, with radar giving high accuracy in estimating biomass under cloudy conditions and laser soundings giving a three-dimensional picture of the forest (Skutsch et al. 2007). Several radar products using SAR such as CAOSMO-SkyMEd, Radarsat-2, TerraSAR-X and L, TanDEM-X and RiSat-1 are being developed. The limitation of these techniques is the expertise required for the analysis and the cost of the data, but the techniques hold great promise for the future.

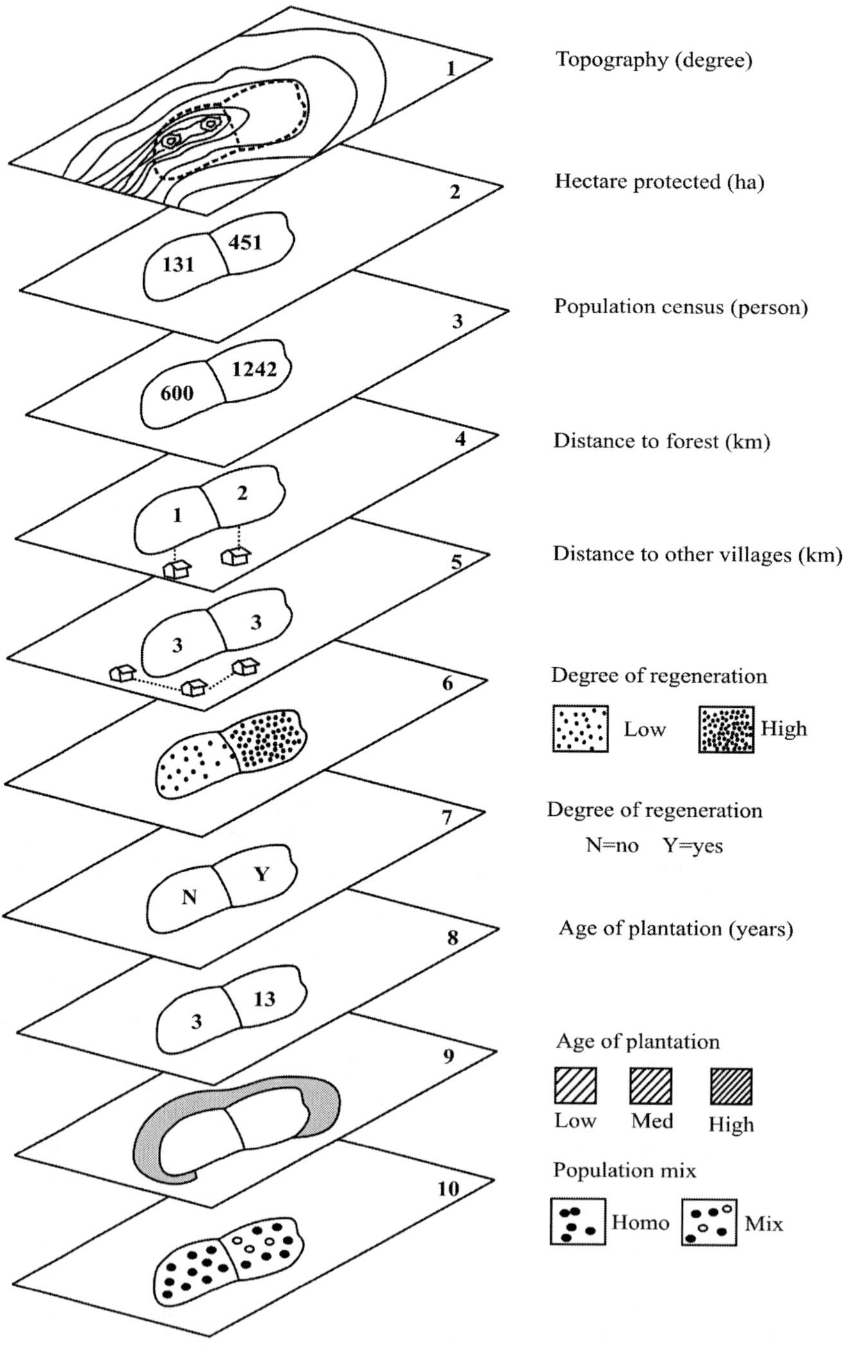

Fig. 14.4 Illustration of the structure in GIS

14.8 Conclusions

Remote sensing and GIS can be useful to most types of land-use project but certain characteristics should be taken in to account when using it. Remote sensing and GIS could be used to

- Define land area and boundary, since GIS allows for good data storages and handling that can be handy throughout the different phases of the project.
- Detect distinct land-use categories and its boundary, but land-use categories with similar vegetation features or blurry boundaries are harder to assess accurately.
- Estimate land cover and land-use change since data can span effectively over time and remote sensing can compute consistent classification for change detections. However, certain land-use practices such as selective felling in forests is hard to detect with remote sensing.
- Estimate homogeneous vegetation such as single species plantations with high certainty, but, heterogeneous vegetation such as tropical natural forest have larger uncertainties.
- Estimate carbon stock if the remote sensed data is accompanied with biomass data to establish relationships.
- Assess land-use over large areas at low cost. Small areas can also be assessed depending on resolution of data. However, its interpretation should always be used together with ground-truthing.

Remote sensing techniques are already used extensively for monitoring changes in land use and land cover in most countries, largely at macro-levels such as global, national and regional, but their application in monitoring changes in land use, biomass production and carbon stocks at the project level is still evolving. In principle, remote sensing and GIS can be used for estimating changes in area and carbon stocks, even at micro-level. However, large-scale application requires lower costs and more reliable estimates. Remote sensing in combination with GIS will be increasingly used in the years to come for national greenhouse gas inventory programmes as well as land-based carbon mitigation and roundwood production projects.

Chapter 15
Modelling for Estimation and Projection of Carbon Stocks in Land-Use Systems

Models are simplified versions of a system used to estimate and project certain features or functions or outputs of a system. In order to study a system scientifically a set of assumptions about how it works is often made. These assumptions, which usually take the form of mathematical or logical relationships, constitute a model (Law and Kelton 2000). Models are used to make projections of carbon stocks in forests, plantations, grasslands and cropping systems. Models are used to make separate projections for biomass and soil carbon stocks in different pools. Further, models are also available to project above-ground and below-ground biomass separately. Models are often based on several assumptions about data and quantitative relationship between input variables and output values. Thus, model outputs are often characterized by uncertainty due to assumptions made about the relationships between variables.

Why modelling is necessary If the relationships that make up a model are simple enough, it may be possible to use mathematical methods such as algebra, calculus or probability theory to obtain exact information on questions of interest; this is called an analytical solution. However, most real-world systems are too complex to allow realistic models to be evaluated analytically, and these models are studied by the means of simulations. Models can be used for making projections of future carbon stocks in carbon mitigation, roundwood production or grassland development projects and are particularly useful in making projections of changes in carbon stocks at the project development stage; models are also useful in estimating carbon stocks based on measurements for land categories or project activities when data are limited. For example, by using only a single parameter such as diameter at breast height (DBH), which can be easily measured, it is possible to estimate the standing biomass of trees in a forest or plantation at any given point. Process-based models are also used in estimating changes or gain and loss in carbon stocks as part of a national greenhouse gas inventory and, in particular, a carbon inventory.

15.1 Types of Models and Application in Estimating and Projecting Carbon Stocks

Several types of models are used in estimating changes in carbon stocks and in growth rates. These models vary in data requirements, process adopted, outputs generated and their application. In general, all the following models can be used in determining the stocks in or growth rates of carbon pools. The models already in use for such purpose are listed below and their features and applications are presented in this chapter:

(i) Biomass equations (or regression models) for biomass and soil carbon projection
(ii) PROCOMAP for project-level carbon stock projection
(iii) CO_2FIX for estimating biomass and changes in soil carbon stock
(iv) CENTURY and ROTH for dynamics of soil carbon

The features, outputs and their applications are summarized in Table 15.1. The details of data needed and steps in adopting the models are described in the following sections.

15.1.1 Biomass Equations

Features Biomass estimation equations are also known as allometric equations or regression models. Normally, these models estimate the biomass or volume of above-ground tree components (kg/tree) based on DBH and height data. These equations are derived based on measured values of tree weight related to its DBH and height from sample trees. Regression models are also available for estimating biomass (tonnes) on a per hectare basis, based on estimates of basal area (m^2/ha) of all tree stems derived using the DBH values of sample trees which are extrapolated to per hectare value. The method for deriving biomass regression equations is given in Chapter 17. The DBH alone or DBH and height may not always explain all the variation in the weight of the tree, the dependent variable. The suitability of a regression model is explained by the standard error of the regression coefficients and the coefficient of the determination (r^2), normally given along with the equation. Normally, each biomass equation can be used within a defined DBH range, for example, a DBH-based equation derived using only large trees (e.g. DBH greater than 30 cm) cannot be used for young trees with low DBH (e.g. smaller than 10 cm). However, most equations can be used for trees within the normal range of DBH, but not for those outside it.

Application Regression models are available for estimating the stocks of different carbon pools, with separate equations for each. These models can be used for estimating the following:

Table 15.1 Comparative features, inputs, outputs and application of carbon estimation and projection models

Model	Features	Key inputs	Key outputs	Application
Biomass equations	- Biomass stock at a given point is estimated using tree parameters	Tree parameters - DBH - Height - Basal area	- Biomass stock estimates kg/tree at a given period - Total AGB and BGB stock/ha	- Afforestation, avoided deforestation, roundwood production
PROCOMAP	- Equilibrium model for estimating carbon stocks for project area	- Area dedicated to activity - Planting rate and vegetation carbon stock in base year - Rotation period - Mean annual increment in biomass and soil	- Total carbon stock/ha and total project area - Biomass and soil carbon stock - Incremental carbon stocks - Cost-effectiveness	- Projection of carbon stocks in forestry mitigation, afforestation, reforestation, avoided deforestation projects
CO_2FIX	- Simulates carbon dynamics of single/multiple species, forest stands with trees of varied ages and agroforestry systems	- Simulation length - Max. biomass in stand - Carbon content - Wood density - Initial carbon - Yield tables - Temperature - Precipitation - LGP	- C stocks and fluxes - Total biomass and soil carbon - AGB and BGB, deadwood, litter and SOC production or stocks	- Forestry and plantation projects - Projection of C stocks in forestry projects for selected species
CENTURY	- Simulates long-term dynamics of C, N, P and S for different plant soil systems	- Monthly mean maximum and minimum air temperature and total precipitation - Plant N, P and S content - Soil texture - Atmospheric and soil N inputs - Initial soil carbon, N, P and S levels	*Soil outputs* - Total C - Soil water dynamics *Plant outputs* - Commercial crop yield - Total dry matter products - Carbon input in plant debris	- Forest, grassland, savannah and cropping systems or projects - Can be applied at plot, project, regional or national level

C: carbon, N: nitrogen, P: phosphorus, S: sulphur, AGB: above-ground biomass, BGB: below-ground biomass, SOC: soil organic carbon, LGP: length of the growing period (season)

- Above-ground biomass (kg/tree) based on DBH or DBH and height
- Above-ground biomass (t/ha) based on basal area (m^2/ha)
- Below-ground biomass (t/ha) based on above-ground biomass

The above-ground biomass equations based on DBH variable are the most commonly used biomass equations. The application of biomass equations along with examples is demonstrated in Chapter 17. These models can be used for estimating carbon stocks or changes in carbon stocks in biomass and in soil for

- Afforestation and reforestation projects under Clean Development Mechanism (CDM) of the Kyoto Protocol

- Industrial roundwood or bioenergy plantation projects
- Community forestry and agroforestry projects
- National greenhouse gas inventory

15.1.2 PROCOMAP

Features PROCOMAP (project comprehensive mitigation analysis process) is a set of models developed by LBNL (Lawrence Berkeley National Laboratory) aimed at estimating the quantity of carbon sequestration achieved for a given year or over a period of years as well as the financial implications and cost-effectiveness of forestry mitigation projects (Sathaye and Meyers 1995). These models use linear growth rates for biomass and soil carbon increments. A PROCOMAP model estimates the following under baseline and mitigation scenario

- Changes in C stock (biomass and soil) annually and cumulatively
  - tC/ha and for the total area
- Cost-effectiveness indicators such as
  - Cost in $/tC sequestered
  - Cost in $/ha
  - NPV (net present value) in $/tC sequestered or emission avoided

Data input for the model includes changes in area under forests and degraded lands under baseline scenario, area proposed for afforestation or reforestation under mitigation scenario, carbon densities of vegetation and soil, rates of carbon sequestration, costs and benefits.

Application A number of variants of PROCOMAP are available for assessing various types of forestry mitigation projects or options, such as

- Afforestation/reforestation: short rotation, long-rotation and natural regeneration
- Avoided deforestation
- Improved forest management
- Bioenergy

These models have been extensively used for assessing mitigation potential (Sathaye and Ravindranath 1998; Ravindranath and Sathaye 2002; Murthy et al. 2006; Ravindranath et al. 2007). These models may require some modification for use in CDM project methodologies and, although not used so far, can be modified for estimating production of industrial roundwood or in traditional forestry projects as well.

15.1.3 Co_2FIX

Features The CO_2FIX model is developed under the Carbon Sequestration and Sustainable Forest Management (CASFOR) project as an inter-institutional collaboration involving ALTERRA, the Netherlands; The Instituto de Ecologia of University of Mexico, Mexico; The Centro Agronomico Tropical de Investigacion y Ensenanza (CATIE), Costa Rica and European Forest Institute, Finland. The version CO_2FIX V 3.1 is a simple carbon bookkeeping model that consists of six modules, one each for the following:

- Biomass
- Soil
- Wood products
- Bioenergy
- Carbon accounting
- Financial

The biomass module converts volumetric net annual increment data with the help of additional parameters, described in the following sections, to annual carbon stocks in the biomass pool. In the soil module, decomposition of litter and harvest residues is simulated using basic information on climate and litter quality. The fate of the harvested carbon is determined in the wood products module, using parameters like processing efficiency, product longevity and recycling. In the bioenergy module, discarded products or by-products from the product module can be used to generate bioenergy, using different technologies. The carbon accounting module keeps track of all fluxes to and from the atmosphere and determines the effects of the chosen scenarios using different approaches to carbon accounting. The financial module uses costs of and revenues from management interventions to determine the financial profitability of different scenarios.

CO_2FIX is an easy-to-use model, which simulates stocks and fluxes of carbon in trees, soil, and – in case of a managed forest – the wood products as well as costs and revenues and carbon credits that can be earned under different accounting systems. Stocks, fluxes, costs, revenues and carbon credits are simulated on a hectare scale with a time step of 1 year. The basic input into the model is the volume of stem growth (from the yield tables or allometric equations or research studies) and parameters on the allocation of that volume or biomass to other plant components such as leaves, branches and roots. Carbon stocks in the living biomass are calculated as the balance between growth and loss (from turnover, mortality and harvest).

Application The CO_2FIX model can be used for estimating the following outputs for a single species or multiple species, forest or plantation and agroforestry systems:

- Carbon stock changes in above-ground biomass
- Carbon stock changes in below-ground biomass
- Carbon stock changes in deadwood and litter
- Changes in stock of organic carbon in soil

The model also provides outputs on wood products as well as financial parameters and can be used for estimating changes in stocks of carbon in biomass and soil for

- Forest sector mitigation projects such as afforestation and reforestation
- Bioenergy mitigation projects
- Plantations of industrial roundwood or fuelwood
- Estimation of carbon credits in CDM projects
- Carbon inventory as a component of the national greenhouse gas inventory

The model is available free of charge (more detailed information on CO_2FIX model is available at http://www.efi.fi/projects/casfor/).

15.1.4 CENTURY

Features CENTURY simulates the long-term dynamics of carbon (C), nitrogen (N), phosphorus (P) and sulphur (S) in different plant–soil systems (http://www.nrel.colostate.edu/projects/century/). The model can simulate the dynamics of grassland, cropland, forest land, and savannah systems. The grassland/crop and forest systems have different plant production submodels that are linked to a common SOM (soil organic matter) submodel. The savannah submodel uses the grassland/crop and forest subsystems and allows for the two subsystems to interact through shading effects and nitrogen competition. The SOM submodel simulates the flow of C, N, P and S through plant litter and the different inorganic and organic pools in the soil. The model runs using a time step of 1 month. The major input variables for CENTURY are presented in Section 15.2.4. These input variables are available for most natural and agricultural ecosystems and can generally be estimated from existing literature.

Applications CENTURY can be used for estimating plant production, yields of commercial crops, carbon input to soil as plant debris and soil organic matter for the following types of ecosystems or vegetation types

- Forest ecosystems
- Grazed ecosystems
- Arable ecosystems

The model can be applied for estimation of total carbon and biomass carbon at plot, project and national level.

Outputs The outputs of CENTURY model include values related to plant and soil such as

- *Soil outputs* total carbon, soil water dynamics and soil temperature dynamics
- *Plant outputs* yield of commercial crops, total dry matter products, and carbon input in soil in plant debris

15.1.5 ROTH3C-26.3

Features ROTH models are developed at the Rothmstead Agricultural Research Station, UK (http://www.rothamsted.bbsrc.ac.uk/aen/carbon/rothc.htm). ROTH C is a model for estimating the turnover of organic carbon in topsoils that takes into account the effects of soil type, temperature, moisture content and plant cover on the turnover process. The model uses a monthly time step to calculate total organic carbon (t/ha) and microbial biomass carbon (t/ha) on a years-to-centuries timescale; it needs few inputs and those it needs are easily obtainable.

Application The main outputs of the ROTH model used for different land-use systems are

- Total organic carbon content in tC/ha in the topsoil
- Carbon content in microbial biomass in tC/ha in the topsoil (optional)

The model can be used for estimating the organic carbon content of grassland, forest land and cropland ecosystems.

15.1.6 Application at National Level for Greenhouse Gas Inventory

The models described above in Sections 15.1.1–15.1.5 can also be used at the national level for estimating biomass and soil carbon stocks and changes for different land categories and subcategories such as forest land (eucalyptus plantation or deciduous forest), grassland and cropland. These models are yet to be adopted for national greenhouse gas inventories. Adoption of these models at national level would require multistage stratification of land categories at the national level to that of homogenous strata, and input data need to be generated and given for these strata.

15.2 Description of Models, Data Needs and Procedure

15.2.1 Steps in Applying Biomass Equations

Methods for and steps in developing biomass equations are described in Chapter 17. Further, the application of biomass equations was also described in Chapter 9. A summary of the steps is provided here:

Step 1: Select the biomass equation relevant to the region, forest type, plantation species and age of the stand

$$\text{Biomass (tropical wet forests)} = 21.297 - 6.953 \times \text{DBH} + 0.740 \times \text{DBH}^2$$

(Brown 1997)

$$\text{Teak volume in m}^3 \ \text{V} = -0.001384 + 0.363126 \ \text{DBH}^2 \ \text{H}$$

(FSI 1996)

Step 2: Tabulate and enter the tree number, DBH and height (H) data into a computer data analysis package such as Excel for each sample plot
Step 3: Enter the biomass equation in the data file or worksheet for estimating the weight of the individual tree (kg/tree) for a given DBH
Step 4: Estimate and add the weight of each tree derived using the DBH values from the sample plot selected
Step 5: Add up the values of total weight of the trees in each of the sample plot to obtain the total weight of all the trees in the sample plots selected for the land category, subcategory and stratum (such as a high-density irrigated eucalyptus plantation)
Step 6: Extrapolate the biomass of trees from the sampled area to per hectare (t/ha)
Step 7: If the biomass equation estimates volume (in m^3), convert the volume into biomass by multiplying the volume by the wood density of the dominant tree species in the plantation or forest.

The steps to be adopted for different pools are identical to the above steps described for estimating above-ground biomass using DBH values.

15.2.2 Steps in Applying PROCOMAP

There are several modules of PROCOMAP for different types of mitigation projects, as described earlier in the chapter. Here, an example of one of the mitigation project types, namely reforestation, is used to demonstrate the steps involved in estimating the carbon sequestration in a short-rotation or long-rotation forest or plantation project. The model estimates the carbon stock changes in the baseline scenario, mitigation scenario and the incremental carbon stocks.

Step 1: Define land use categories

- Define land categories relevant to baseline as well as mitigation scenario, for example, degraded land, forest land and plantation categories

Step 2: Define baseline area under different land categories

- Select the baseline land category and define the area for a selected base year (e.g. 2007) and project the area under this category annually for future years up to, say, 2037
- If any projections are available for the project area or region or national level for the land category selected, use them

- If no projections are available, make projections using demographic, social and economic factors
- Note that the degraded land area is normally taken to remain stable or increase over the years whereas forest area is assumed to decline because of anthropogenic pressures

Step 3: Estimate the area under reforestation options

- Select the reforestation option such as short-rotation or long-rotation forest plantation
- Enter area to be reforested yearly (e.g. from 2007 to 2037, which could be constant or vary from year to year

Step 4: Estimate carbon densities in soil and vegetation under baseline scenario

- Enter the carbon densities of vegetation (above-ground woody biomass) and soil in tC/ha
- Add soil carbon density and vegetation carbon density to get total carbon density/hectare

Step 5: Calculate carbon density under the mitigation scenario

(i) *Vegetation or biomass*

- Enter rotation period, carbon density (normally 0.5 tC/t of woody biomass) and rate of carbon accumulation (tC/ha/year) in vegetation under the mitigation scenario for the activities chosen
- The rate of carbon accumulation depends on a number of factors such as tree species, density, rainfall, nutrient supplements and rotation period
- Define rotation period, which could vary for different reforestation options
  - Short-rotation forestry: 5–10 years
  - Long-rotation forestry (for sawn wood): 30–50 years
  - Carbon sequestration storage projects: duration indefinite

(ii) *Soil*

- Enter rate of carbon accumulation in soil carbon for the activities selected (tC/ha/year) under reforestation option, which involve planting trees, soil carbon density increases due to litter fall and decomposition

(iii) *Carbon in decomposing matter*

- Enter the carbo3n density in decomposing matter (tC/ha) and the decomposing period

(iv) *Carbon in wood products*

- Enter the carbon stock in the product pool (tC/ha) and the average age of retention of carbon in the products
- The woody biomass sequestered and harvested has diverse end-uses, where carbon emissions occur at different periods

- Potential uses are
  - As fuelwood, where combustion leads to carbon emission instantly
  - As industrial roundwood (for pulp production), where emissions occur over 2–5 years
  - As sawn wood for long-term storage, where emissions occur after 30–50 years

Step 6: Generate outputs of the PROCOMAP REFOREST module

15.2.3 Steps in Applying CO_2FIX (Version 3.1.0)

Establishing the initial parameters in CO_2FIX model A CO_2FIX file needs to be prepared for each species or group of species with supporting parameters. Parameterization is an important step in estimating carbon stocks. Initial parameters need to be specified for each module. The growth of stem biomass (Bs) is expressed as a function of age. The values are initially derived from yield tables and further refined with data from inventories. The biomass allocation coefficients (F) for foliage, branches and roots are expressed as a function of the age of the tree. Like the growth rates of the stem, the F-parameters have to be refined through destructive sampling. The data used to determine the parameter values are based on scientific and peer-reviewed literature. The CO_2FIX model description and its user manual give the data sources and references to research studies on which the model parameters were developed.

Parameterization The following aspects need to be considered in parameterization of the CO_2FIX model:

- The CO_2FIX manual should be consulted for implementing the model and for its parameterization.
- The factors influencing carbon stocks can be captured using the data and parameters from yield tables, local studies, official publications and peer-reviewed literature on vegetation, soil and climate of the region.
- It is a good practice to assess the mean, median and range for each parameter required in the CO_2FIX model from local studies and published literature relevant to the region.
- Multiple model runs should be done to calibrate the parameters and the model projections compared with actual data on carbon pools to assess the robustness of the parameters.

Estimation of changes in the carbon stock of tree biomass Most yield tables report only merchantable stem volume and exclude information on the branch and leaf biomass. In such situations, the CO_2FIX model could be used to estimate the changes in total above-ground tree biomass:

- Estimates of stem volume from yield tables or other studies to be collected and incorporated under the species files in the CO_2FIX model.

- The projections of CO_2FIX model are in the time steps of 1 year and follow the "Gain–Loss" approach, which takes into account growth and loss of tree biomass during the year from harvest and disturbances such as fire and pests. The annual change in the tree carbon stock is reflected in the *ex ante* estimation (pre-implementation phase). The growth parameters reflect the annual increases and losses in above-ground tree biomass due to thinning, harvests and disturbance
- The harvested biomass from thinning and harvest is subtracted from the existing biomass, and the slash and deadwood from the harvests are added to the soil module since these are expected to decompose over time.

Estimation of carbon stock changes in non-tree shrub biomass The shrub biomass can be estimated by modelling the non-tree woody perennial species as a cohort of species. The data on shrubs can be collected from local studies or published literature and used for parameterizing shrub growth in the model to estimate shrub biomass and projected change in its stock of carbon.

Estimation of carbon stock changes in below-ground biomass In the CO_2FIX model, the relation between below-ground biomass and above-ground biomass is expressed as a fraction of stem biomass for each species.

Estimation of carbon stock changes in deadwood In the CO_2FIX model, deadwood is included in coarse woody litter (stems and stumps) and, to a lesser extent, in the short fine woody litter (fine and coarse branches, coarse roots).

Estimation of carbon stock changes in litter The data on estimates of litter from literature can be used to parameterize the CO_2FIX model. Data on litter can be input either directly into the model or estimated using the biomass module through biomass turnover, natural mortality, management mortality and logging slash.

Estimation of changes in soil organic carbon CO_2FIX uses the Yasso model for modelling soil carbon dynamics. The Yasso model (see, e.g. Liski et al. 2005) describes the decomposition and dynamics of soil carbon and calibrates the total stock of soil carbon without distinguishing between soil layers. The model uses parameters from the soil module, deadwood and litter parameters from the biomass module and climate parameter inputs under the general parameters tab. The soil module consists of two tabs, namely general parameters and cohort parameters. The parameters for the soil module are under the soil main menu. The user needs to provide climate parameters for the site. The model has been tested in evaluating the effects of climate on decomposition rates of litter in a wide range of ecosystems.

Steps in using the graphical user interface CO_2FIX V. 3.1.0 Model The following broad steps can be adopted to enter data and to generate outputs:

Step 1: Open a new project by clicking the "New" option in the file menu
Step 2: Go to the data menu and scroll down to general parameters; define the project scenario and enter the general parameters
Step 3: Select the data menu and scroll down to biomass; as you click, a new window appears; enter the input parameters related to biomass parameters

Step 4: In the data menu, go to "Soil" and define general and cohort parameters
Step 5: Define the products, bioenergy, finance and carbon credit sections if these are relevant to the project goals
Step 6: Go to "View" and get any of the following outputs

- Carbon stock
- Financial
- Carbon credit

Step 7: Go to the "File" menu and export the output file to a desired location and in the desired format

15.2.4 Steps in Applying CENTURY (V. 5)

The information required by the CENTURY model can be considered under two categories, namely:

- Site characteristics data, which is related to the type of land mapping unit or land facet and the type of ecosystem sustained by such land unit
- Data on variables necessary for the parameterization of the model for the selected ecosystem: forest, grassland and agricultural land

Data input The following data are required for the model, which are site specific for the region and ecosystem selected:

- Mean monthly precipitation
- Maximum and minimum temperature
- Content of lignin in the plant material
- Content of N, P and S in the plant material
- Texture of the soil
- Initial contents of total S, C, N and P in the soil
- Schedule of agricultural, livestock or forestry activities
- Levels and amounts of agricultural inputs used during the management cycle
- Production by land use types (crop yields or other outputs in tonnes per hectare)
- Soil erosion (as soil losses in kg/m^2)
- Date of disturbance (e.g. incidence of fire, clearings or other disturbing phenomena during the study period)

Parameterization of the model The first step is to define the major land category type, namely forest land, grassland and agricultural land, and the subsystem or project activity. Enter the specific information needed to parameterize the model for each ecosystem or project activity according to the modules.

Module 1 Bulk density, number of soil layers or soil profile horizons, drainage pattern, permanent wilting point of the soil (lower limit of soil moisture for plant growth) and field capacity of the soil (moisture left is soil a few days after rain or irrigation) and pH.

Module 2 Labile organic C (g/m^2), non-labile organic C (g/m^2), C/N ratio per soil layer, initial inputs of plant residues (g/m^2), C/N ratio of litter in soil, C/N ratio of the soil organic horizon and value of the C isotope in land cover (litter) (g/cm^2).

Module 3 Amount of C in foliage in the forest system, amount of N in foliage in the forest system, amount of C in fine and coarse branches, amount of N in fine and coarse branches, amount of C in fine and coarse roots, amount of N in fine and coarse roots and initial amount of C in dead material.

Linking data file structures and the CENTURY programme module CENTURY model consists of a large number of files as default values in the model, which can be used for generating outputs since these values have been generated and standardized for a range of ecosystems.

The model generates detailed outputs related to carbon, nitrogen and other components. The variables relevant for carbon sequestration are those that are related to fluxes of carbon in soil and CO_2 release.

Steps in using the graphical user interface for CENTURY (V.5) model The following steps could be used for operationalization and running of the model:

Step 1: Open the CENTURY graphical user interface (GUI); a window displaying "Session history" and "Messages" appears along with "Quick steps" buttons, namely "Preferences" "Sites" "Management" "Output file" "Status" and "Run simulation"

Step 2: Go to "Preferences" and set the project directories

Step 3: Create a new site parameter or edit an existing one; the site parameters are listed above, under "Input"

Step 4: Go to "Management" and specify the site management parameters by specifying general simulation information and defining the management block(s)

Step 5: Go to "Output file" and select the output file format: choose either of the two output formats, namely NetCDF or spreadsheet (ASCII); NetCDF is more useful for gridded data and spreadsheet is more convenient to handle non-gridded data

Step 6: Run the simulation

Step 7: Go to the "Result" menu and view/plot/browse/export the simulation results

15.2.5 Steps in Applying ROTH

To run ROTH, it is necessary to prepare a series of input files that contain climate, soil and land management information. The running of the model involves the following steps:

Step 1: Input climate data such as monthly mean temperature, total monthly precipitation and total monthly open pan evaporation

Step 2: Input soil data such as percentage of clay in soil and soil depth
Step 3: Input land management data related to carbon simulation

- Monthly inputs of organic matter to soil
- Plant residues (tC/ha) added monthly
- Farmyard manure (tC/ha) added monthly
- Soil cover (covered/fallow)

Step 4: *Parameterize the model* by running the model itself using varied additions of organic matter input to soil to achieve equilibrium

- Values of soil organic carbon present in soil

Step 5: Run the model for parameterization

- Select the time period

Step 6: Run the Roth C model for generating soil carbon dynamics
Step 7: Note the soil carbon dynamics outputs

- Carbon dynamics outputs under "current conditions scenario" of soil, climate and management for the selected time period
- Carbon dynamics outputs under the "new management scenario" such as a mitigation scenario

15.3 Conclusions

The models described in this chapter could be applied for projecting changes in carbon stock at the project or national level. The models are particularly required for projecting changes in carbon stocks or roundwood production during the project development phase as well as during the earlier periods of post-implementation phase. The selection of the model depends on the following:

- Objective of the programme, such as

  (i) Estimation or projection of changes in carbon stock due to mitigation project activities
  (ii) Estimation of carbon emissions and removals for greenhouse gas inventory
  (iii) Estimation or projection of industrial roundwood or fuelwood
  (iv) Understanding the carbon dynamics

- Input data availability for the model
- Accuracy required
- Access to model and suitability of the model to the location, land category or project activity

Among the models described in this chapter, biomass estimation equations are extensively used for most mitigation and roundwood production projects as well as for estimations related to national greenhouse gas inventories. Models will be increasingly used for mitigation projects, greenhouse gas inventories and round-wood production programmes. The use of models is becoming increasingly important and some of the models are being recommended for afforestation and reforestation projects under the CDM of the Kyoto Protocol. Very often, the application of models is limited by the availability of input data. Care should be taken in selecting the model, since all models have some limitations due to the assumptions and the relationships developed between various input and output variables.

Chapter 16
Carbon Inventory Methods for National Greenhouse Gas Inventory

All countries that are signatories to the United Nations Framework Convention on Climate Change (UNFCCC) are expected to prepare national greenhouse gas (GHG) inventories, annually in the case of Annex-1 (industrialized) countries and periodically – once in 3–5 years – in the case of Non-annex-1 (largely the developing) countries. A national GHG inventory requires estimation of GHG emissions by source and removals by sinks from all sectors in a country for a given year or period. All the member countries of the UNFCCC are expected to use the *Revised 1996 Guidelines* for National GHG Inventory for Land Use Change and Forest Sector (LUCF). The use of *Revised 1996 Guidelines* for GHG inventory is set out in various decisions and conclusions of the UNFCCC (IPCC 2002). *The Revised 1996 IPCC Guidelines* stipulate the following sectors for national GHG inventory:

1. Energy
2. Industrial Processes
3. Agriculture
4. Land-Use Change and Forestry (LUCF)
5. Waste

Further, IPCC prepared Good Practice Guidance (IPCC 2003) for Land-Use Land Use Change and Forest Sector (LULUCF) to reduce uncertainty in the inventory estimates. IPCC (2000) defines inventories consistent with good practice as those which contain neither overestimates nor underestimates, so far as can be judged, and in which uncertainties are reduced so far as is practicable. The Good Practice Guidance (IPCC 2003) provides guidelines for carbon inventories for afforestation, reforestation and avoided deforestation projects in addition to those for land-use categories such as forest land, cropland and grassland.

Countries preparing the GHG inventory for submission to the UNFCCC use *The Revised 1996 IPCC Guidelines* along with the Good Practice Guidance. However, IPCC has further prepared *IPCC 2006 GHG Inventory Guidelines* based on the experience gained in using *The Revised 1996 IPCC Guidelines* and improvements in scientific methods for the following sectors:

- Energy
- Industrial Processes and Products Use
- Agriculture, Forests and Other Land-uses (AFOLU)
- Waste

Each sector comprises categories and subcategories. Guidelines for the AFOLU sector of *IPCC 2006 Guideline* attempts to overcome many of the limitations of the earlier IPCC 1996 Guidelines by providing the following:

- *Good practice guidance for all GHGs* for all land-use categories and carbon pools
- *Tiers*, which represent differing levels of complexity in methods
- *Key category analysis* to identify categories that have significant influence on the total inventory for the country
- *Default data* (improved default data to enable use of new the inventory guidelines)
- *Worksheets* to calculate emissions and removals
- *Uncertainty estimation* methods and guidance on reducing uncertainty
- *Quality assurance (QA) and quality control (QC)* to check and review the estimates for ensuring quality of the estimates
- *Reporting framework* to report the national inventory estimates

The inventory guideline also provides guidance on ensuring quality at all stages of inventory compilation and aims at enhancing transparency, completeness, consistency, comparability and accuracy. The IPCC 2006 Inventory Guidelines are available for countries to adopt in their preparation of national GHG inventories.

Definitions of terms Some of the terms used in the IPCC GHG inventory guidelines are source, sink, activity data, emission factor and removal factor.

Source Any process or activity that releases a GHG (such as CO_2 and CH_4) into the atmosphere. A carbon pool can be a source of carbon to the atmosphere if less carbon is flowing into it than is flowing out of it.

Sink Any process, activity or mechanism that removes a GHG from the atmosphere. A given pool can be a sink for atmospheric carbon if during a given time interval more carbon flows into it than flows out of it.

Activity data (AD) Data on the magnitude of human activity resulting in emissions/ removals taking place during a given period of time (e.g. data on land area, area afforested, area converted to another land-use and roundwood extraction).

Emission factor (EF) A coefficient that relates the activity data to the amount of chemical compound that is the source of emissions. Emission factors are often based on a sample of measurement data, averaged to develop a representative rate of emission for a given activity level under a given set of operating conditions (e.g. above-ground biomass stock per hectare or soil organic carbon density).

Removal factor Rate at which carbon is taken up from the atmosphere by terrestrial systems such as forests and grassland and sequestered in biomass and soil (e.g. above-ground biomass growth rate).

16.1 The Revised 1996 IPCC Guidelines

The Revised 1996 IPCC Guidelines have been extensively used by over 100 countries in preparing the national GHG inventory. The GHG inventories submitted have been compiled and synthesized by UNFCCC. The methodological issues largely relate to the following:

- Lack of compatibility between IPCC land/forest category/vegetation type/systems/formats and national circumstances or classification in most countries
- High uncertainty in inventory estimation
- Lack of clarity in reporting estimates of emissions/removals in managed natural forest
- Lack of consistency in estimating/reporting total biomass or only above-ground biomass
- Reporting total biomass covering either multiple carbon pools or a single pool, such as above-ground biomass; lack of consistency for comparability
- Guidance for below-ground biomass not provided
- Estimation (or differentiation) of managed (anthropogenically impacted) and natural forests
- Lack of methods for savannah/grassland
- Lack of methods for incorporating non-forest areas such as coffee, tea, coconut and cashew nut, as well as ambiguity about agroforestry
- Absence of linkage between biomass and soil carbon
    - In *Revised 1996 IPCC Guideline* changes in stocks of biomass and soil carbon are estimated in different IPCC categories or worksheets and are not linked

16.2 The IPCC 2003 and 2006 Guidelines

The fundamental basis of the methodology of inventory rests upon two linked assumptions: (i) the flux of CO_2 to/from the atmosphere is equal to changes in carbon stocks in the existing biomass and soils and (ii) changes in carbon stocks can be estimated by first establishing the rates of change in land-use and the practice used to bring about the change (e.g. burning, clear-cutting, selective cutting, change in silviculture or other management practice). This requires estimation of land use in the inventory year, conversion of forest land or grasslands, and the stocks of carbon in the land-use categories (those that are subjected to change and those that are not).

The unique feature of land-use sector is that it includes emissions as well as removals, unlike in other sectors where only GHG emissions are estimated. This sector encompasses all land-use categories of a country. The key improvements in the IPCC (2003, 2006) Guidelines with respect to GHG inventory for land-use categories over 1996 that are relevant to carbon inventory are summarized below:

- *Adoption of six land-use categories* forest land, cropland, grassland, wetland, settlements and other land to make the estimates consistent by covering all land-use categories over the years. These land-use categories are further disaggregated to account for the carbon dynamics, especially in the soil, due to land-use change into
  - *Land remaining in the same category* such as forest land remaining forest or grassland remaining grassland (involving no change in land use).
  - *Land converted to another land-use category* such as grassland or cropland converted to forest land.
- *Adoption of key source or sink category analysis* for different land-use categories, CO_2 pools and non-CO_2 gases to focus on the dominant land-use categories, GHGs and carbon pools.
- *Adoption of three hierarchical tier methods* that range from default emission factors and simple equations to the use of country-specific data and models to accommodate national circumstances (see Section 16.8 for details).
- *Linking biomass and soil carbon* in all land-use categories such as forests and croplands.
- *Guidance for all the five carbon pools* namely, above-ground biomass, below-ground biomass, deadwood, litter and soil organic carbon.
- *Provision of generic methods* to account for changes in stocks of carbon in biomass and soil in different land-use categories.

The IPCC (2003, 2006) provides guidance on the following steps for preparing the inventory:

1. *Definition of managed land* Divide all land into managed and unmanaged land, since GHG inventory is made only for lands subjected to anthropogenic effects, which, in other words, is managed land.
2. *Land classification* Develop a national land classification system applicable to all six land-use categories (forest land, cropland, grassland, wetland, settlements and other land) and further subdivide by climate, soil type and/or ecological regions appropriate for the country.
3. *Compilation of data* Compile data on the area and changes in the area in each land-use category and subcategory if available. Categorize land area by specific management systems defined for each land-use category, and by subcategory, if available. This categorization could provide the basis for assigning emission factors and stock-change factors required for calculation of CO_2 emissions and removals.
4. *Estimation of CO_2 emissions and removal* Estimate emissions and removals at the appropriate tier level based on key category analysis.
5. *Estimation of uncertainties and adoption of quality assurance and quality control (QA/QC) procedures* Use methods provided to estimate uncertainty.
6. *Calculation of total inventory* Sum CO_2 emissions and removals over the inventory period for each land-use category and subcategory.
7. *Reporting of inventory* Using reporting tables, convert carbon stock changes to net emission or removals of CO_2 by land-use categories.

8. *Documentation and archiving* Document and archive all information used to produce an inventory, including activity and other input data, emission factors, sources of data and data documentation, methods, description of model and QA/QC procedures and reports in addition to the results for each source category.

16.3 Carbon Inventory Methods for Land-Use Categories

Estimation of CO_2 emissions and removals is the dominant component of GHG Inventory Guidelines. The basic principle in estimating CO_2 emissions or removals from land-use categories involves combining the information on the extent to which human activity takes place (called activity data or AD) with the coefficients that quantify the emissions or removals per unit of the activity. These coefficients are called emission factors (EF) or removal factors (RF). Examples of activity data are the area of forest or plantations harvested or planted or supplied with fertilizers, and examples of emission or removal factors include growth rate of above-ground biomass, density of soil carbon and quantity of biomass burnt or harvested. The basic equation is as follows:

$$CO_2 \text{ emission and removal} = \text{AD} \times \text{EF or RF}$$

Guidance is provided for estimating CO_2 emissions and removals from all land-use categories under two conditions due to differing soil carbon dynamics:

- *Land remaining in the same land-use category* such as forest land remaining forest land
- *Land converted to other land-use category* such as grassland converted to forest land

Two methods are provided for carbon inventory, namely: (i) 'Gain–Loss' method and (ii) "Stock-Difference" method. Carbon inventory involves estimating the net emissions and removals for each land-use category for a given year, aggregated to all land-use categories.

16.4 What IPCC 2003 and 2006 Inventory Guidelines Provide

The inventory guidelines provide guidance separately for each of the carbon pools and for each of the land-use categories and also separately for the two subcategories, namely: (i) land remaining in the same category and (ii) land converted to other land-use category. The inventory guidelines provide the following information for each carbon pool separately.

- *Choice of method* The approach to be adopted is discussed, particularly the equations to be used for calculating the carbon gain–loss for a given year or stock at different periods. The approach and methods are described for the three tiers, particularly in detail for Tier 1 (refer to Section 16.8).
- *Choice of emission or removal factors* The approach to selecting emission or removal factors is presented for the three Tiers.
- *Choice of activity data* Approach to selecting the activity data is presented according to the three Tiers.
- *Calculation steps* Brief steps in estimating carbon gain–loss or stock changes involving activity data and emission or removal factors are given by using a set of equations.
- *Uncertainty assessment* The extent of and approach to estimating uncertainty in the emission or removal factors, activity data and the total estimate of CO_2 emissions and removals are covered.

16.5 What IPCC 1996, 2003 and 2006 Inventory Guidelines do not Provide

The IPCC guidelines are meant to assist carbon inventory teams at the national level in calculating carbon emissions and removals and do not provide guidance on the following:

- *Methods for estimating carbon stocks* Methods for measuring or monitoring carbon in biomass or soil from different land-use categories are not given. The examples of methods are harvest and permanent plot method. The frequency of measurement of changes in carbon pools is not described.
- *Methods for estimating growth rates* Methods are not described for measurement and calculation of growth rates of biomass and soil carbon pools.
- *Sampling methods and procedures* Methods of selecting and locating the type, size and number of plots for measurement of carbon pools are not provided.
- *Methods for determination of area* Methods for estimating the area under a given land-use and marking of the boundary for a given land-use management practice are not given.
- *Parameters to be measured* The parameters to be measured or monitored in the field or laboratory for calculating biomass and soil carbon are not given. Examples of parameters are tree DBH (diameter at breast height) and height, soil bulk density, soil organic carbon content and woody biomass density.
- *Field and laboratory measurement techniques* Methods and techniques for measuring the parameters for estimating biomass and soil carbon are not described. Examples of techniques include those for measuring tree height and DBH, soil bulk density and wood density.
- *Calculation procedures* Procedure for conversion of parameters (such as height, DBH, soil bulk density) measured into biomass or soil organic carbon density per hectare is not given.

- *Models for carbon estimation* Estimation or projection of biomass stock changes using models such as CO_2FIX, PROCOMAP and CENTURY is not described.

Many countries have forest inventory programmes for monitoring biomass stocks and changes in those stocks. Data from such forest inventories could be used for the national carbon inventory. Guidance on preparing a forest inventory, particularly for above-ground biomass and commercial or merchantable timber is available in literature (Kangas and Maltamo 2006). However, the traditional forest inventory may have the following limitations for the national carbon inventory.

- Cropland, agroforestry and grassland categories are not included in the forest inventory
- Normally, forest inventory provides estimates of above-ground biomass, particularly commercial or merchantable timber, and it may not include estimates for soil organic carbon, below-ground biomass, deadwood and litter pools
- Non-tree biomass is not estimated even for forest lands
- Land-use changes or boundary shifts are not included

Therefore, there is a need for guidance on the methods of estimating activity data and emission factors to improve the estimates included in the national GHG inventory.

16.6 Application of Carbon Inventory Methods to National GHG Inventory

Activity data and emissions and removals factors are critical to a reliable and high-quality GHG inventory. The key activity data on area under different forest and plantation types, cropping systems, grassland and so on are collected in most countries for several purposes of planning and development and thus are often available, although data on changes in area are often limited. Even FAO and other international agencies publish data on the activity data. Thus, availability and quality of activity data are better than emission/removal factors for most countries.

However, the national GHG inventory process, particularly in the developing or tropical countries, is characterized by non-availability of emission and removal factors for different land-use categories, subcategories and land management systems. Even when available, the uncertainty level of the data is high, which makes the estimates in the emission inventory incomplete or characterized by high uncertainty. Majority of the developing countries use a large number of default carbon emission and removal factors provided in the IPCC guidelines, the Emission Factor Data Base (EFDB), FAO publications and other global data sources. The emission and removal factors are highly variable, often location-specific and likely to vary with soil, rainfall, altitude and management systems, even for a given forest type, plantation species, grassland or cropland systems.

Uncertainty in the estimates that make up the carbon inventory is likely to be high when global default values for parameters such as growth rates of above-ground biomass, biomass and soil carbon stocks are used. Developing a database on carbon stocks and growth rates of different carbon pools for various land-use categories (forest, cropland and grassland) and subcategories based on soil, rainfall and other agroclimatic conditions and management systems at subnational level can reduce such uncertainty.

The methods and guidelines provided in this handbook can be used for generating emission and removal factors required for estimating carbon gain–loss or stock changes at the national level for the GHG inventory. Nationally derived emission or removal factors for different land-use categories, subcategories and management systems at disaggregated level will enable a country to reduce uncertainty in the estimates of national GHG inventory. The methods provided in the earlier chapters could be adopted for generating emission and removal factors for different forest and plantation types, grassland management, cropland and agroforestry systems. Some of the key emission and removal factors required for the national CO_2 inventory are

(i) Above-ground biomass stock and growth rates
(ii) Below-ground biomass stocks
(iii) Deadwood and litter stocks
(iv) Soil organic carbon stocks and rates of change
(v) Density of woody biomass
(vi) Biomass conversion and expansion factors

16.7 Approach to Generating Carbon Emission and Removal Factors for National GHG Inventories

Preparing a GHG inventory is a long-term and recurring process under the UNFCCC, which is why it is important for all countries to have long-term and definitive institutional arrangements for the process, particularly for land-use sectors. The following approach could be adopted for using carbon inventory methods to improve national GHG inventories.

16.7.1 Tier Definition and Selection for GHG Inventory

A tier represents a level of methodological complexity. A shift from lower tier to higher tier is associated with increased complexity, data requirements and accuracy, ultimately reducing uncertainty of the estimates. IPCC (2003, 2006) inventory guidelines provide three tiers, which implicitly progress from the least to the highest level of certainty in estimates as a function of

- *Methodological complexity*
- *Regional specificity of model parameters*
- *Spatial resolution and extent of activity data*

Tier 1 Tier 1 is the simplest and basic method to use, requiring minimal data. The activity data and emission factors, particularly the latter, are obtained from global data sources, such as FAO reports and websites. National-level activity data are often available but the availability of emission and removal factors is limited. The global or regional data sources are often in the form of aggregated or macro-level averages. Countries using Tier 1 methods usually use the national sources of activity data and default emission and removal factors from IPCC guidelines or the FAO database. Inventory estimates made using Tier 1 methods are associated with high uncertainty.
Tier 2 Tier 2 uses the same methods or equations as those used in Tier 1, but the activity data and emission factors are derived from national sources. Tier 2 can use the same methodological approach as Tier 1 but applies emission and stock change factors that are based on country- or region-specific data for the most important land-use categories. Country-defined emission factors are more appropriate for the climatic regions, land-use systems and management systems in the given country. Higher temporal and spatial resolution and more disaggregated activity data are typically used in Tier 2 to correspond to country-defined coefficients for specific regions and specialized land-use systems.
Tier 3 Higher-order methods are used in Tier 3, including models and inventory measurement systems tailored to address national circumstances, repeated over time, and driven by high-resolution activity data disaggregated at subnational level. These higher-order methods provide estimates of greater certainty than those obtained from the lower tiers. Such systems may include comprehensive field sampling repeated regularly and/or GIS-based systems of age class, production data, soils data and land-use and management activity data integrating several types of monitoring. A piece of land where a land-use change occurs can usually be tracked over time using remote sensing techniques. Models should undergo quality checks, audits and validations and be thoroughly documented.

Combining the tiers Although it is desirable to adopt higher tiers for all key land-use categories, subcategories and carbon pools, it may not be feasible due to data and resource limitations. The next best approach is to adopt a combination of tiers for increasing the accuracy of the inventory estimates. For example, Tier 2 could be adopted for activity data on a forest plantation and Tier 1 default estimates of mean annual above-ground biomass growth rate for the same land-use category. Further, Tier 2 estimates, based on location-specific studies, can be used for biomass stock and Tier 1 estimates for soil carbon stock for a given forest or plantation category. A combination of tiers, with preference for higher tiers, could be adopted

- *For different land-use categories*
- *Within a given land-use category for different carbon pools*
- *Within a carbon pool, for activity data and emission factors*

16.7.2 Key Category Analysis

The concept of key category analysis is used to identify the key land-use categories and subcategories that have a significant impact on a country's total carbon inventory from the land-use (AFOLU) categories. The key category analysis is conducted at multiple levels. *Firstly*, to select the key sector among the IPCC sectors, namely energy, industrial processes and product use, agriculture, AFOLU (or LULUCF) and waste. *Secondly*, among the key sectors selected within AFOLU or LULUCF, to select key land-use categories, for example, forest land or grassland or cropland. The key category analysis can be extended to identify key land subcategories also. *Thirdly*, to select carbon pools. In this section the focus is only on AFOLU or LULUCF sector. The key category analysis could be used to identify the dominant carbon pools that impact the carbon inventory for a given

- *Land-use category*: forest land or grassland or cropland
- *Subcategory*: forest land remaining forest land, cropland converted to forest land, grassland converted to cropland, cropland converted to forest land, etc.
- *GHGs*: CO_2, CH_4 and N_2O

Key category analysis would assist a country in efficient allocation of limited resources to key categories for the selection of method, size of the sample, frequency of monitoring a given carbon pool, data collection and analysis. By focusing on the dominant land-use categories or carbon pools, key category analysis will minimize the uncertainty.

If the key category analysis shows that a certain land-use category or carbon pool is a key source or sink, it is necessary to adopt Tier 2 or Tier 3. This requires estimation of emission or removal factors at the national and subnational level according to rainfall zones, soil types, management systems, etc. The approach to conducting key category analysis is as follows (IPCC 2003, 2006).

- *Selection of level of analysis* The analysis should be performed at the level of IPCC categories (e.g. forest land or grassland) or subcategories (e.g. forest land remaining forest land, land converted to forest land or grassland remaining grassland). A country may adopt further disaggregation depending on the availability of activity data and emission factors, for example
    - *Forest land can be further disaggregated into* evergreen forest, deciduous forest, or eucalyptus plantations
    - *Carbon pools can be further disaggregated into* above-ground biomass, below-ground biomass, deadwood, litter and soil organic carbon
- *Method of analysis* The key categories are identified using a previous inventory estimate for the sector and a predetermined cumulative emissions threshold. The cumulative emissions threshold is normally 95% of the total level of emissions or removals (IPCC 2006). The method of analysis could be at different levels namely

- *Land-use* category where the emissions or removals are aggregated for each land-use category, including subcategories, to select the key land-use category or subcategory
- *Carbon pools* where the emissions from all carbon pools are aggregated to select the key pool for a selected key land-use category

- *Procedure for calculating the contribution of each land-use category or carbon pool* The following equation could be used for estimating the contribution

Key category assessment = land-use category or carbon pool estimate/total contribution of all land-use categories or carbon pools

- *Ranking the land-use categories or carbon pools* Using the estimates of contribution of land-use categories, subcategories and carbon pools are ranked and the key categories selected using the threshold

16.7.3 Land-Use Categories and Stratification

Using the guidelines provided by the IPCC (IPCC 2003, 2006), the geographic area of the country is categorized according to

- *Land-use categories* forest land, cropland, grassland, etc.
- *Subcategories* land remaining in the same use category and land converted to a new land-use category
- *Vegetation and management systems*
  - *Forest land* evergreen forest, moist deciduous forest, dry forest, mountain systems, scrubland and tropical desert
  - *Plantations* Eucalyptus, Teak, Acacia, Pine, etc., age of the stand (<5 years, 5–10 years, 10–20 years, >20 years)
  - *Cropland* irrigated, rainfed, agroforestry systems, annual and perennial crops, area under rice, sorghum, maize, etc.
  - *Grassland* intensively managed grassland and open-access grazing land, fertilized or irrigated grassland
- *Rainfall zones*
  - Humid, subhumid, semi-arid and arid
  - Wet, moist and dry zones
- *Other criteria* such as soil type, altitude and slope

Disaggregate the geographic area of the country according to the above or some other nationally relevant criteria into homogeneous categories, subcategories or land-use and management systems. Disaggregation or stratification of land-use systems into finer and homogeneous systems will reduce the error or uncertainty in using the carbon stocks and rates of change of different pools for national estimates. However, finer levels of disaggregation add to the cost of inventory estimation and are subject to data limitations, particularly the emission and removal factors.

16.7.4 Selection of Land-Use Categories, Estimation of Area and Preparation of Spatial Maps of the Land-Use Categories

Land-use stratification may lead to a large number of forest or plantation types with differing management systems, a large diversity of cropping systems, etc. Stratification is very critical since the area under different land strata could be large, covering millions of hectares, with diverse rainfall, soil and topographic conditions. Thus, multistage stratification may be needed. Select the land-use strata for monitoring based on criteria such as extent of area under land-use system, distribution of area under a given land-use system into areas under different rainfall, soil and management systems. It may not be feasible to select all the strata for each land-use

Table 16.1 Stratification of forest and non-forest land-use categories

Land use	Ecological zone	Type of forest	Crown cover (%)	Altitude (m)
Tropical forest	Tropical moist deciduous forest	– Degraded – Secondary – Old growth	–>70 –50–70 –10–50 <10	<500 500–1,000 >1,000
	Tropical dry forest	– Degraded – Secondary – Native	–>70 –50–70 –10–50 –<10	<500 500–1,000 >1,000
	Species	*Rainfall* (cm/year)	*Age group* (years)	*Density of trees*
Plantations	Eucalyptus	50–100 100–200 >200	<5 5–10 10–20 >20	<2,000 2,000–4,000 >4,000
	Pines	50–100 100–200 >200	<5 years 5–10 years 10–20 years >20 years	<2,000 2,000–4,000 >4,000
	Seasonality	*Rainfall* (cm/year)	*Crop system*	*Crops*
Cropland	Annual crops	–50–100 –100–200 –>200	– Irrigated – Rainfed	– Sorghum – Maize – Rice
	Perennial crops	–50–100 –100–200 –>200	– Irrigated – Rainfed	– Coconut – Mango
	Agroforestry system	–50–100 –100–200 –>200	– Irrigated – Rainfed	– Mixed: tree + annual crops – Orchards

system because of constraints of data and financial and other resources. Obtain a land-use map of the country with land-use categories and wherever possible stratify each category into subcategories based on vegetation type, management system and other criteria. Obtain the area under each strata consisting of land-use category, subcategory and any additional criteria. Higher the levels of stratification into homogenous categories, lower will be the error in estimation. Area under different strata and its spatial distribution is necessary for sampling for long-term monitoring. An illustration of land-use stratification, based on IPCC guidelines, is given in Table 16.1.

The national land-use map showing spatial distribution of different strata should be overlaid on a latitude–longitude grid for sampling. If GIS facility is available, different layers of maps, each representing a land-use or management system or rainfall zone or other feature, could be overlaid.

16.8 Estimation and Monitoring of Biomass Stocks and Changes

16.8.1 Sampling Method and Location of the Plots

The sampling method described in Chapter 10 could be adopted to determine the sample size, number and size of plots. The sampling procedure will be more complex since, at a national level, the area covered under each land stratum or category is large and spatially spread over large regions with diverse rainfall, soil and topographic conditions and management systems, compared to projects, which are normally spread over thousands of hectares. Thus, the sample plots could be larger than those of carbon mitigation or land development projects, since the area under each land strata is likely to be large, running to millions of hectares. The selected number of plots should be located on the spatial latitude–longitude grid map depicting different land strata. These sample plots should be located in the field using GPS readings and marked on the field as well as on the map (refer to Chapter 10 for the procedure).

16.8.2 Permanent Plot Method for Biomass Carbon

Long-term monitoring of selected carbon pools in the land-use strata would require adoption of the "permanent plot" technique (refer to Chapter 10). The technique enables periodical measurement and monitoring of carbon pools leading to calculation of rates of growth or decline in a carbon pool. These plots should be located in a way that facilitates periodic visits over a number of years or even decades. The permanent plots must be located using permanent markers and landmarks on the ground as well as backed by the GPS readings. The locations of shrub and herb plots as well as points of soil sampling should be marked on the maps, along with their GPS readings. It is always desirable to have larger plots for adopting the "permanent plot" technique.

16.8.3 Parameters to be Monitored for Biomass Carbon Inventory and Frequency of Monitoring

All the indicator parameters relevant to estimating the stocks of different carbon pools should be considered for periodical monitoring of different land-use category strata (Table 16.2). The selection of carbon pools for each land-use category would depend on the key category analysis. The following parameters could be considered for monitoring different carbon pools. The final selection of the parameters for field monitoring would depend on the land-use category and judgment of inventory experts.

The frequency of monitoring would vary with the carbon pool and the parameter. The frequency of monitoring different pools and parameters is discussed in Chapter 4. Normally, the above-ground biomass and soil carbon are the key pools for carbon inventory of forests, grassland and cropland categories. Above-ground biomass for tree species is monitored once in 2 or 3 years and soil organic carbon is monitored once in 3–5 years. This may be adequate for reporting GHG inventory. GHG inventory is made annually or for a selected year, but the year in which carbon inventory was undertaken for different carbon pools may be different and may require extrapolation or averaging.

Table 16.2 Parameters to be monitored for carbon inventory for different pools

Carbon pool	Vegetation type	Parameters
Above-ground biomass	Trees	Species name DBH and height Crown cover
	Non-tree	Species name Height and DBH, if relevant Density of plants/plot Fresh and dry weight of biomass
Below-ground biomass	Tree and non-tree	Weight of the root biomass in the soil core sample Volume of the core
Deadwood	Tree	Species name, DBH and height of dead and standing tree
Litter	Tree and non-tree	Standing litter weight in sample plots at different periods
Soil organic matter	All land-use categories	Depth of soil sample Bulk density of the soil Fresh weight of the soil sample Organic matter content

16.8.4 Preparation for Fieldwork, Data Formats and Field Measurement Procedures

The material required and preparation for fieldwork for different carbon pools are described in the earlier chapters: above-ground biomass in Chapter 10, below-ground biomass in Chapter 11, deadwood and litter in Chapter 12 and soil organic carbon in Chapter 13. Refer to the respective chapters for the formats for recording field data and measurement procedures and techniques for different carbon pools. Given the national scale of sampling and measurements, it is very important to properly document and archive data for different carbon pools, plots, land-use category and subcategory, management systems and regions within a country. The data could be recorded and stored according to latitude–longitude grids in addition to the above features.

16.8.5 Methods of Analysis and Calculation Procedure for Biomass Carbon

The methods for estimating carbon stocks in different biomass pools using the indicator parameters measured in the field and laboratory are described in Chapter 17. The emission and removal factors required for calculation of CO_2 emissions and removals from different land-use categories are above-ground and below-ground biomass stocks and growth rates, soil organic carbon stocks and growth rates, and deadwood and litter stocks.

The emission and removal factors developed using these guidelines for carbon inventory could be used in estimating CO_2 emissions or removals based on the methods and procedures given in the IPCC guidelines (IPCC 2003 or 2006). The IPCC guidelines provide the worksheets for calculating CO_2 emissions or removals, which work on the principle of emissions being equal to the product of activity data and emission factors.

The worksheets are provided (IPCC 1996, 2003, 2006) for each category and subcategory, as a series of tables where the columns consist of activity data and emission factors.

16.9 Estimation and Monitoring Stocks and Change in Stock of Soil Organic Carbon

The National GHG inventory requires estimates of changes in stocks of soil organic carbon in different land-use categories and subcategories for the inventory year. The methods for estimating soil organic carbon are described in Chapter 13. The approach to estimating stocks of soil organic carbon in different land-use categories is summarized as in the following steps:

Step 1: *Selection of land-use categories and subcategories* The approach and methods for selecting land-use categories and subcategories described for estimating biomass carbon stocks in Chapter 10 should be adopted for soil organic carbon estimation.

Step 2: *Stratification of land-use categories, area estimation and demarcation* The stratification of land-use categories and subcategories adopted for biomass carbon estimation should be adopted for soil organic carbon also, so that the biomass and soil carbon estimate can be linked for the same land area. Different land-use categories, subcategories and strata should be marked on a latitude–longitude map of the country to facilitate periodical visits for sampling.

Step 3: *Sampling method and location of the plots* The permanent plot method suggested for biomass carbon should be adopted for soil carbon. For sampling too, the method adopted for biomass carbon for different land-use categories, subcategories and strata should be followed for soil organic carbon. The approach to selection of sampling points within the plots selected for biomass carbon, described in Chapter 13, could be adopted here also. The sampling points selected for collecting soil samples should be marked on the latitude–longitude grid map of the country along with GPS readings for the location.

Step 4: *Selection of parameters and frequency of monitoring* The parameters to be selected for estimating soil organic carbon are the same as described for projects in Chapter 13, namely

- Soil depth
- Soil bulk density
- Soil organic matter content

The frequency of monitoring of soil organic carbon is usually once in 3–5 years, since the incremental change in the stock is likely to be marginal compared to the stock that exists in the soil.

Step 5: *Field methods, preparation for field sampling, field measurements and formats* The material required for sampling, preparation for fieldwork, soil sample collection and bulk density measurement techniques described in Chapter 13 could be adopted here.

Step 6: *Laboratory estimation method* Several laboratory methods are described in Chapter 13. Among the methods described, the "wet digestion method" could be adopted.

Step 7: *Methods of calculation* The method for calculating tonnes of carbon per hectare based on values of soil depth, soil bulk density and organic matter content (%), described in Chapter 17, could be used for each land-use category, subcategory and stratum during the year of measurement. The soil organic matter density (tonnes of carbon per hectare) should be used in the worksheets along with the data on area for each stratum (IPCC 1996, 2003, 2006).

16.10 Reporting of GHG Inventory Estimates for LULUCF or AFOLU Sector

The basic reporting table for GHG emissions/removals estimates prepared using the IPCC 2003 and 2006 guidelines is shown in Table 16.3. The inventory estimates for reporting in the format shown here can be obtained by adopting the worksheets and compilation sheets provided in IPCC (2003, 2006).

16.11 Uncertainty Estimation and Reduction

IPCC (2003, 2006) suggests estimation of uncertainty for different activity data and emission factors, CO_2 emissions and removal estimates for different land-use categories and aggregate national estimates. Uncertainty estimation methods are described in Chapter 18. IPCC (2003, 2006) guidelines describe two methods for

Table 16.3 Basic reporting table for GHG inventory using IPCC (2003, 2006) in Gg. (From *www.unfccc.org*; GHG Inventory of New Zealand, 2005.)

GHG source and sink categories	Net CO_2 emissions or removals	CH_4	N_2O	NO_x	CO
Total land-use categories	–24,594.33	4.04	0.03	1.00	35.38
A. Forest Land	*–25,513.17*	*2.24*	*0.02*	*0.56*	*19.62*
1. Forest land remaining forest land	–26,767.89	0.98	0.01	0.24	8.56
2. Land converted to forest land	1,254.72	1.26	0.01	0.31	11.05
B. Cropland	*–639.14*	–	–	–	–
1. Cropland remaining cropland	–671.12	–	–	–	–
2. Land converted to cropland	31.98	–	–	–	–
C. Grassland	*706.91*	*1.80*	*0.01*	*0.45*	*15.76*
1. Grassland remaining grassland	–,–,–	1.80	0.01	0.45	15.76
2. Land converted to grassland	706.91	–	–	–	–
D. Wetlands	*0.72*	–	–	–	–
1. Wetlands remaining wetlands	–,–	–	–	–	–
2. Land converted to wetlands	0.72	–	–	–	–
E. Settlements	*97.16*	–	–	–	–
1. Settlements remaining settlements	–	–	–	–	–
2. Land converted to settlements	97.16	–	–	–	–
F. Other land	*38.98*	–	–	–	–
1. Other Land remaining other land		–	–	–	–
2. Land converted to other land	38.98	–	–	–	–

The signs for CO_2 removals are negative (–) and CO_2 emissions are positive (+); CH_4, N_2O, NO_x and CO are the other greenhouse gases estimated for land-use sectors the estimation of combined uncertainties, namely simple propagation of errors and Monte Carlo analysis. Use of either provides insights into how individual categories and carbon pools contribute to uncertainty in total emissions in a given year.

16.12 Quality Assurance and Quality Control

IPCC (2000, 2003) provides definitions and guideline for adopting quality assurance (QA) and quality control (QC) practices, keeping in mind the need to enhance transparency and accuracy of the estimates in GHG inventories (refer to Chapter 18).

General QC procedures The general methods focus on processing, handling, documenting, archiving and reporting procedures. An example of QC activity and procedure involves the following:

- Check the integrity of database files
  - Confirm that the appropriate data processing steps are correctly represented in the database
  - Confirm that the data relationships are correctly represented in the database
  - Ensure that data fields are properly labelled and have the correct design specifications
  - Ensure adequate documentation of database and model structure

CO_2 source/sink category-specific QC procedures General QC checks relate to data processing, handling and reporting, whereas CO_2 source/sink category-specific procedures relate to the key categories. QC procedures are directed at specific types of data used in the methods and require knowledge of

- *CO_2 source/sink category*
- *Type of data available*
- *Parameters associated with emissions/removals*

QC procedures focus on the following types of checks (these are only examples; refer to Chapter 5 of IPCC 2003 for details):

- Check that land areas are properly classified and that no double-counting or omission of land area has occurred
- Check consistency of time series activity data
- Check sampling and extrapolation protocols adopted

QA review procedures Quality assurance requires an expert review to assess the quality of the inventory and to identify areas where improvements are necessary. Preliminary QA procedure involves basic expert peer review by inventory agencies. Apply the review process to all source/sink categories, particularly the key categories. Secondary QA procedure involves, for example, expert peer review of calculations, assumptions and models used.

16.13 Remote Sensing Techniques for National Carbon Inventories

Remote sensing techniques are already used for estimating the activity data relevant to land-use categories and subcategories. Remote sensing techniques are beginning to be used for estimation and monitoring of emission and removal factors such as above-ground biomass, below-ground biomass and soil carbon stocks. The details of and approaches to adopting remote sensing techniques for estimating carbon stock values for different carbon pools are given in Chapter 14, which could be adapted to national carbon inventory processes.

16.14 Conclusions

Nearly, all the countries have prepared carbon and GHG inventories at the national level and submitted them to the UNFCCC (www.unfccc.org). National GHG inventory reports, as well as the compilation and synthesis reports of the UNFCCC, have identified a number of issues and problems in estimating CO_2 or other GHGs for the inventory. Some broad issues identified are as follows:

(i) Lack of clarity in the methods as well as inadequacies of the methods
(ii) Lack of activity data and emission factors
(iii) Low quality or reliability of emission factors
(iv) High uncertainty of emission factors, leading to uncertainty in inventory estimates
(v) Unsuitability of default emission factors to national circumstances

The main limitation highlighted by all the studies relates to lack or low reliability of emission and removal factors. The critical step in preparing a reliable carbon inventory is to generate emission and removal factors nationally for disaggregated homogeneous land-use categories and subcategories. The methods provided in this handbook could be adopted for generating emission and removal factors. All the countries have to establish long-term carbon inventory programmes and permanent plots for monitoring carbon stock changes. The guidelines and methods provided in this chapter could be applied in estimating CO_2 emissions and removals from land-use categories based on the *Revised 1996 IPCC Guidelines* (IPCC 1996), *IPCC Good Practice Guidance* (IPCC 2003) and *2006 IPCC Guidelines for National GHG Inventory* (IPCC 2006).

Chapter 17
Estimation of Carbon Stocks and Changes and Data Sources

The main goal of carbon inventory projects and programmes is to estimate carbon stocks and changes in those stocks annually or at different times. Chapters 10–13 described the methods for measuring and monitoring different indicator parameters from which carbon stocks in different carbon pools can be estimated. The next step is to estimate the carbon stocks and changes, using the parameters measured and monitored in the field and in the laboratory. The analysis and calculation of carbon stocks and changes involve conversion of field and laboratory estimates of various parameters from sample plots, such as diameter at breast height (DBH), height and soil organic carbon content, into tonnes of carbon per hectare per year or over several years using different methods and models. The carbon pools for which the stocks are to be estimated are

(i) Above-ground and below-ground biomass
(ii) Deadwood and litter
(iii) Soil organic carbon

The estimates of carbon stocks and changes are required by different stakeholders for different applications at different periods and frequency intervals (Chapter 3):

- *CDM project developers and implementing agencies* require the estimates annually as well as over a period of years, usually 3–5 years, in terms of tonnes of carbon per hectare and for the project area.
- *Roundwood or industrialwood production managers* require estimates of commercial timber production in tonnes or cubic metres per hectare and for the full plantation at the end of rotation period.
- *Community forestry producers* require production estimates of fuelwood and timber in terms of tonnes or cubic metres per hectare over varying time spans.
- *Tree growers or farmers* need periodical estimates of tonnes of fuelwood and cubic metres of timber per hectare.
- *National greenhouse gas inventory agencies* require estimates of emissions and removals of carbon dioxide, in tonnes, from different land-use categories for a given inventory year or on an annual basis.

This chapter presents methods and procedures for calculating carbon stocks and changes in the stocks of different carbon pools.

17.1 Above-Ground Biomass of Trees

Above-ground biomass of trees includes commercial (or merchantable) timber and total tree biomass, which includes not only commercial timber, but also twigs, branches and bark, expressed as tonnes of oven-dried biomass. The field methods for estimating above-ground biomass are described in Chapter 10.

17.1.1 Estimating Tree Biomass Using the Harvest Method

The harvest method was described in Chapter 10 and involves the following steps:

- Selecting sample plots for each stratum (see Section 10.3 for stratum definition) representing a land-use category or project activity
- Harvesting of all the trees in each of the sample plots
- Measuring the fresh weight of the commercial stem, twigs and branches separately for all the harvested trees in the sample plots
- Calculating the dry weight of commercial stem and twigs and branches per plot
- Aggregating the weight of dry tree components from all the sample plots
- Extrapolating tree biomass (stems, twigs and branches) from plot values to tonnes per hectare separately and as total

The calculation procedure is simple, involving, as presented in steps above, extrapolation of the dry weights of all the trees harvested and weighed from sample plots to per hectare and expressed in dry matter terms.

17.1.2 Estimating Tree Volume

The plot method described in Chapter 10 provides values for tree parameters such as DBH and height. These values could be used to estimate the volume of the tress, which can be converted into weight terms using wood density. This method involves the following steps:

Step 1: Measure the height and DBH of all the trees in the sample plots (following Chapter 10).

Step 2: Tabulate the values of height and DBH by species and by plot.

Step 3: Estimate the volume of each tree in the sample plot using the following formulae depending on the shape of the tree, whether cylindrical or conical.

$$V = \pi \times r^2 \times H \quad \text{(for cylindrical shape)}$$

$$V = (\pi \times r^2 \times H)/3 \text{ (for conical shape)}$$

where:

V = volume of the tree in cubic centimetres or cubic metres

r = radius of the tree 130 cm above the ground = DBH/2
H = height of the tree in centimetres or metres.

Step 4: Obtain the wood density value for each of the tree species from literature (refer to Section 17.7), at least for the dominant species

- If the density value for any dominant tree species is not available in the literature, select the species most closely related to the species present in the site.

Step 5: Multiply the volume of the tree with the respective wood density to obtain the dry weight of that tree and convert the weight from grams to kilograms or tonnes.

- Weight of tree (in g) = volume of the tree (in cm^3) × density (g/cm^3)

Step 6: Sum the weight of all trees of each species in the selected sample plots (in kilograms or tonnes for each species).
Step 7: Sum the weight of all the trees of all tree species for all the sample plots, based on the weight calculated for each plot (in kilograms or tonnes).
Step 8: Extrapolate the weight of each species from the total sample area (sum of all the plots) to per hectare value (tonnes of biomass per hectare for each species).
Step 9: Sum the biomass of each species to obtain the total biomass of all the trees in tonnes per hectare (dry matter).

17.1.2.1 Additional Steps for Estimating the Biomass Using Volume

Estimation of wood density The density of wood varies with species, age and even component of tree biomass (main trunk, branches and tender twigs). Thus, it is a good practice to estimate the density of the tree species present in the selected plots, especially if literature values are not available for that species. The following steps could be adopted to estimate the density of wood:

Step 1: Select the tree species and components for which density has to be estimated
Step 2: Extract a piece of wood by drilling or by cutting; collect three to five such pieces
Step 3: Take the fresh weight of each piece of wood (in grams)
Step 4: Dry the wood sample in an oven at 105°C to constant weight
Step 5: Take a graduated measuring jar of known dimensions and volume, of a size convenient for immersing the piece of wood using a pointed needle, to measure the amount of water displaced
Step 6: Record the height of the water and then immerse the piece of wood into water in the measuring jar and measure the increased height of the water in the jar and ensure that the piece of wood is completely submerged in water
Step 7: Estimate the volume of the piece of wood in cubic centimetres (cm^3), based on the height of the water raised and the radius of the jar

$$V = \pi r^2 H$$

where:

V = volume of the piece of wood in cubic centimetres
r = radius of the jar in centimetres
H = increase in the height of the column of water on immersing the piece of wood in centimetres

Step 8: estimate the density of the wood (D) using the following formula:

D (in g/cm^3) = weight of the piece of the dry wood (in g)/ volume of the piece of wood (in cm^3)

17.1.3 Estimating Tree Biomass Using Volume Tables

Volume tables are available for a large number of commercial tree species. A volume table is a two-way table, which gives the volume of a tree if its DBH and height are known. Once volume is known, biomass can be estimated using wood density of the species in question. Table 17.1 can be used to calculate the total biomass per plot and per hectare. The steps provided in Section 17.2.1 can also be followed for estimating the weight of trees and biomass per hectare. In this section, the volume of a tree is estimated using the volume tables. However, these tables provide only the merchantable volume; to convert it to total biomass, it is necessary to use biomass conversion and expansion factors available in literature (refer to Section

Table 17.1 A volume table for *Cedrus deodara*. (From Chaturvedi and Khanna 1982.)

	Volume (m^3)				
	Height (m)				
DBH (cm)	17	23	29	35	41
15	0.030	0.068	0.106		
20	0.114	0.182	0.250		
25	0.222	0.329	0.435	0.541	
30	0.355	0.508	0.661	0.814	
35	0.512	0.720	0.929	1.137	1.345
40	0.692	0.965	1.237	1.509	1.782
45	0.897	1.242	1.586	1.931	2.276
50		1.552	1.977	2.403	2.828
55		1.894	2.409	2.924	3.438
60		2.269	2.882	3.494	4.107
65		2.677	3.396	4.115	4.834
70		3.117	3.951	4.785	5.619
75			4.547	5.504	6.462
80			5.185	6.274	7.363
85			5.863	7.092	8.322
90			6.583	7.961	9.339
95			7.343	8.879	10.415
100			8.145	9.847	11.549

17.1.6). The approach based on volume tables is being increasingly replaced by biomass equations method (Section 17.1.5).

17.1.4 Estimating Tree Biomass Using Mean Tree Weight

Estimating tree biomass involves estimating the mean weight of trees according to DBH classes, multiplying it by the number of trees in the respective DBH classes and adding up the totals of all the classes in a given sample plot to obtain the total weight of all the trees in the plot, which is then extrapolated to per hectare by taking into account all the sample plots. Estimating the weight in this manner avoids felling a large number of trees and the associated cost of estimating their weight. The tree biomass method is thus an attractive option, although it is not as accurate as using species-specific biomass equations derived from weighing felled trees. The method involves the following steps:

Step 1: Select the forest or tree plantation stratum, select sample plots and measure the DBH of all the trees in the sample plots

Step 2: Using DBH data from field measurements in sample plots, prepare a frequency table using appropriate class interval of DBH for each tree species

- To reduce error, it is desirable to have smaller DBH class intervals (e.g. 5 cm or smaller) depending on the range of the DBH in the sample plots

Step 3: Tabulate the DBH data (Table 17.2 illustrates the format)

Step 4: Extrapolate the total number of trees of the selected species from sample plots to per hectare values to fill that column in the table

Step 5: Locate a tree with DBH closest to the mean DBH value in the forest or plantation for each DBH class for the selected tree species

Step 6: Harvest the selected tree with the mean DBH and estimate the fresh weight of the mean tree

- If required, separate the harvested tree biomass into commercial stem, twigs and branches and estimate the fresh weight separately for each

Step 7: Estimate the dry weight of the selected tree for each DBH class by selecting a sample of fresh wood for drying in the hot air oven at 105°C, to constant weight; if needed, sample each component separately and add up the total

Step 8: Estimate the total weight of all the trees in a hectare in each DBH class, using the dry weight of the tree with mean DBH and the number of trees in that DBH class

Step 9: Aggregate the biomass of all trees present in a hectare

Step 10: Repeat the steps for each species and aggregate the weight of all the tree species on per hectare basis

- For tree species sparsely represented in the sample plots, use the mean weights derived for other species which has similar tree structure (bole and crown size and shape)

Table 17.2 Estimating above-ground biomass of *Syzigium cuminii* from DBH values

DBH class (cm)	Mean DBH from sample plots (cm)	Mean dry weight of tree with mean DBH (kg/tree)	No. of trees/ha	Total biomass, dry weight (kg/ha)
5–10	8.0	15	5	75
10–15	12.5	21	25	525
15–20	18.0	28	175	4,900
20–25	24.0	40	200	8,000
25–30	28.0	55	175	9,625
>30	33.0	70	100	7,000
Total			680	30,125

Application The mean tree method is a practical method, which can be used in locations and for tree species for which biomass estimation equations are either not available or cannot be determined.

17.1.5 Biomass Equations

Biomass equations can be used to estimate the weight of the tree based on the measured DBH or DBH and height of each tree in the sample plots. Using biomass equations is a common and cost-effective method to estimate biomass of tree species present in a forest or plantation, and has been explained in detail in Chapter 15.

This section illustrates how the equations are developed and applied in estimating the weight of individual trees as well as biomass per hectare of forests and plantations.

Developing a biomass equation Biomass equations are available only for some dominant commercial tree species. A few generic equations for broad forest types are also available. However, the equations are often specific to not only species, but also locations. Further, biomass equations developed using mature trees cannot be used for younger trees, and vice versa. Biomass equations are not available for most local or native forest tree species, which makes it desirable to develop biomass equations wherever possible to suit the local tree species and age of the stand. The process involves the following steps:

Step 1: Select the forest or plantation strata, select the sample plots, identify the dominant tree species and measure the DBH and height of all the trees in the sample plots (Chapter 10)

Step 2: Select a few dominant tree species from the forest or plantation

Step 3: For the identified tree species, randomly select at least 30 trees to represent different girth sizes present in the forest or plantation

Step 4: Measure DBH and height of each tree of the selected species

Step 5: Cut the selected trees to the ground and separate each felled tree into the trunk, twigs and branches

Step 6: Cut the tree into pieces of an appropriate size to weigh them directly with a field balance and determine dry weight

- If cutting a large tree trunk into smaller pieces is not feasible, estimate the volume ($\pi r^2 H$) of a large piece of the tree trunk by measuring its diameter and height; r = (diameter/2) and H = height

Step 7: Estimate the weight of each tree by weighing the whole tree; for large trees, calculate the weight by multiplying the volume with wood density

Step 8: Plot the weight of the tree (y-axis) against the DBH (x-axis) to obtain the best fit between the variables; the curve could be linear or non-linear

Step 9: Develop the biomass equation linking tree weight to DBH alone or to DBH and height using any of the software packages available, such as Microsoft Excel or SPSS (statistical package for social scientists)

- Methods for developing linear and non-linear biomass equations using data on DBH, height and weight of trees are given in most textbooks on statistics and forest mensuration. Further discussion on developing and using biomass equations can be found in Brown (1997) and Parresol (1999)
- Estimate the constant (a) and regression coefficient (b) for the biomass equation along with the coefficient of determination (r^2), which explains the proportion or percent variation in biomass (dependent variable) as a function of DBH (independent variable)

One of the limitations of biomass equations is that harvesting about 30 or more trees of a given tree species may not be feasible or permitted, except for plantation species. Further, the harvesting process is also expensive. Therefore, biomass equations are developed only for the dominant tree species; for the rest, generic equations can be used.

Application of biomass equation Most projects of climate change mitigation and roundwood production and greenhouse gas inventories require estimation of above-ground biomass, and it can be estimated as follows:

Step 1: Select the biomass equation for the dominant species present in the forest or plantation

- The equation may have been developed for the specific project or for another project in the region with similar vegetation, soil and other features
- If locally developed biomass equation is not available, search for one in the literature or in databases developed for a species with similar tree structure
- If species-specific biomass equations are not available, use generic equations

Step 2: Enter the DBH and height data of measured trees in the sample plots separately for each tree species into a database

Step 3: Enter the equation into Excel or any other appropriate software and link the database to the equations

Step 4: Use DBH and height data along with species-specific biomass equation to estimate the weight of individual trees using a software package

The weight of the tree can be estimated even using a simple calculator by substituting the DBH (and height) into the equation, especially if the number of observations is small

Step 5: Aggregate the biomass of all trees in the sample plots for each species for which biomass equation is developed or available

Step 6: Extrapolate the biomass values (tonnes) from sample plots to per hectare biomass value (tonnes dry biomass per hectare)

Step 7: If biomass equation is not available for some minor tree species, use the biomass equation of a tree species closest to it in terms of tree form, height and crown spread

Biomass conversion and expansion factor In using biomass equations developed using only the merchantable volume, it is necessary to convert the value to whole-tree biomass (which includes stems, twigs and branches) using Biomass Conversion and Expansion Factors (Table 17.3). These factors can be generated in the field or obtained from literature (for details, refer to Section 17.1.6).

Biomass equations based on DBH and/or height Biomass equations can be linear, quadratic, cubic, logarithmic and exponential. Often generic biomass equations are used for estimating the above-ground biomass (Table 17.4). In addition to biomass equations for individual trees, they are also available for estimating biomass at per hectare scale.

Usually only the volume of a tree is measured since measuring the weight, particularly of large trees, is difficult in the field. Many biomass equations are indeed biomass volume equations. Tree volume is related to parameters such as DBH and height. The volume (m^3) estimated using the equations needs to be converted to biomass in tonnes per tree or per hectare using the density of the species. The following steps could be adopted for estimating the volume as well as the biomass of the trees:

Step 1: Select the forest or plantation strata, select sample plots, identify the dominant tree species and measure the DBH and height of all the trees in the sample plots

Step 2: Select the biomass volume estimation equation for the dominant tree species or for all the species for which species-specific equations are available (Table 17.5)

Table 17.3 Default biomass conversion and expansion factors (BCEF) (tonnes of biomass per cubic metre of wood volume) for humid tropical zone. (From IPCC 2006.)

Growing stock level (m^3)							
<10	11–20	21–40	41–60	61–80	80–120	120–200	>200
Conifers							
4	1.75	1.25	1	0.8	0.76	0.7	0.7
(3–6)	(1.4–2.4)	(1–1.5)	(0.8–1.2)	(0.7–1.2)	(0.6–1)	(0.6–0.9)	(0.6–0.9)
Natural forest							
9	4	2.8	2.05	1.7	1.5	1.3	0.95
(4–12)	(2.5–4.5)	(1.4–3.4)	(1.2–2.5)	(1.2–2.2)	(1–1.8)	(0.9–1.6)	(0.7–1.1)

Table 17.4 Some generic equations for estimating biomass

Forest type[a]	Equation	R^2/sample size	DBH range (cm)
Tropical moist hardwoods[b]	$Y = \text{EXP}[-2.289 + 2.694\text{LN(DBH)} -0.021(\text{LN(DBH)})]$	0.98/226	5–148
Tropical wet hardwoods[b]	$Y = 21.297 - 6.953(\text{DBH}) + 0.740(\text{DBH})$	0.92/176	4–112
Temperate/tropical pines	$Y = 0.887 + [(10{,}486(\text{DBH})^{2.84}/(\text{DBH}^{2.84}) + 376{,}907]$	0.98/137	0.6–56
Temperate US eastern hardwoods	$Y = 0.5 + [(25{,}000(\text{DBH})^{2.5})/((\text{DBH}^{2.5}) + 246{,}872$	0.99/454	1.3–83.2

[a]Updated from Brown (1997), Brown and Schroeder (1999) and Schroeder et al. (1997); Y = dry biomass in kg/tree, DBH = diameter at breast height, LN = natural log; EXP = "e raised to the power of"

[b]Delaney et al. (1999); Y = biomass in kg/tree, HT = height of trunk (m), LN = natural log

Table 17.5 Some species-specific biomass equations based on GBH (girth at breast height) values. (From Kale et al. 2004)

Species	Model	a	b	R^2	SE
Bauhinia racemosa	$Y = a + b^*X$ ($X = \text{GBH}^{2*}\text{height}$)	0.0431	0.0025	0.97	3.17
Zizphus xylopyra	$\log_{10} Y = a + b^*\log X$ (X = GBH)	−3.20	2.87	0.94	0.12
Tectona grandis	$\text{Log } Y = a + b^*\log X$ (X = GBH)	−2.85	2.655	0.98	0.075
Lannea coromandelica	$\text{Y} = \text{a} + \text{b}^*\text{X}$ ($X = \text{GBH}^{2*}\text{height}$)	−1.84	0.002	0.98	14.49
Miliusa tomentosa	$Y = a + b^*X$ ($X = \text{GBH}^{2*}\text{height}$)	−0.68	0.0024	0.99	1.33

- If no species-specific equations are available, use generic equations or those specific to a given forest or plantation type

Step 3: Enter the DBH, height and the biomass volume equation into a software package such as Excel

Step 4: Calculate the volume of each tree based on the DBH or DBH and height using the software

Step 5: Aggregate the volume of all the sample trees by species if species-specific equations are used to obtain the total volume of the trees (m^3)

Step 6: Convert the volume of the trees in the sample plots to biomass in tonnes using density of biomass for the selected species

- If species-specific density values are not available or cannot be derived for all the species, use the density of the dominant species for converting the whole forest or plantation volume to biomass

- If the equation provides only the merchantable volume, use biomass conversion and expansion factor to obtain total biomass in kilograms per hectare or tonnes per hectare

Step 7: Extrapolate the biomass from the sample plot area to tonnes of biomass per hectare.

Basal area based biomass equations Basal area of all the trees of a given species or of all trees in the sample plots can be estimated using the DBH values.

$$\text{Basal area } (\text{m}^2/\text{ha}) = \sum \pi \ (\text{DBH}/2)^2$$

$$\text{Evergreen forest: biomass (t/ha)} = -2.81 + 6.78\ (\text{BA}),\ r^2 = 0.53\ \text{(Murali et al. 2005)}$$

$$\text{Deciduous forest: biomass (t/ha)} = 11.27 + 6.03\ (\text{BA}) + 1.83\ (\text{height}),\ r^2 = 0.94\ \text{(Murali et al. 2005)}$$

$$\text{Eucalyptus hybrid volume } (V \text{ in m}^3/\text{ha}) = -2.4992 + 0.287\ (\text{BA}) + \text{height (from field measurements)}$$

where:

BA = basal area in square metre per hectare.

17.1.6 Biomass Conversion and Expansion Factors

The biomass volume tables and the default biomass stock as well as growth rates are often estimated considering only the merchantable or commercial volume. Estimating only the commercial component of the tree biomass, which is largely the main tree trunk, may be adequate for industrial roundwood producers. However, for estimating carbon stocks and changes for climate change mitigation projects and greenhouse gas inventory estimation, all the above-ground biomass including twigs and branches and even leaves needs to be estimated. To convert merchantable tree volume into total biomass, biomass conversion and expansion factors (BCEF) are used (IPCC 2006). Biomass conversion and expansion factors expand dry weight of the merchantable volume of above-ground biomass or growing stock to account for non-merchantable components of the tree, stand and forest. Such factors can also be developed for local tree species by harvesting trees and estimating the proportion of merchantable and non-merchantable components. Default BCEF values are also available in literature (IPCC 2006). The factors can be estimated using the following broad steps at a species level or for a given forest type:

Step 1: Select tree species or the forest type for determining BCEF
Step 2: Adopt the sampling method described for estimating above-ground biomass from the harvest method
Step 3: Select at least 30 trees of a given species or a mixture of species for a selected forest type

Step 4: Cut the trees to the base and segregate the above-ground tree into the main tree trunk or the merchantable bole and the remaining tree components such as tapering tree stems, branches and twigs

Step 5: Estimate the volume of the merchantable tree component and the remaining biomass using methods described for the harvest method (Section 17.1.1) and extrapolate the estimate obtained from the trees and sample plots to a per hectare value.

Step 6: Estimate BCEF by calculating the ratio (BEF) of total volume to merchantable volume and multiplying by density:

$$\text{BEF} = (\text{Total volume of trees/ha})/(\text{merchantable volume of trees/ha})$$

$$\text{BCEF (in t/m}^3) = \text{BEF} \times \text{density (t/m}^3)$$

BCEF can be estimated for a given species, a forest type consisting of multiple tree species and young or mature plantations or forests.

Biomass expansion factors (BEF) could be used if a biomass equation provides the merchantable biomass (t/ha) directly. BEF expands the dry weight of the merchantable volume of the growing stock to account for non-merchantable components of trees.

Total biomass can be estimated in two ways depending on the units of merchantable biomass estimates (as volume in cubic metres or in tonnes per hectare).

$$\text{Total biomass (t/ha)} = (\text{total merchantable biomass (t/ha)}) \times \text{BEF}$$

$$\text{Total biomass (t/ha)} = \text{merchantable biomass in volume (m}^3\text{/ha)} \times \text{BCE}$$

17.1.7 Application of Methods for Estimating Above-Ground Biomass

Different methods described for estimating above-ground biomass are being used by different stakeholders. Among all the methods, biomass (including volume) estimation equations are likely to be applied the most. However, the number of species for which such equations are available is limited. In the absence of species-specific biomass equations, generic biomass equations are frequently used. Thus, it may be desirable for project developers and inventory experts to develop region-specific and species-specific equations. Once developed, they can be used for other projects in the region. In the absence of biomass equations, the simplest method is to estimate the volume of sample trees using DBH, height and the form factor and then converting the tree volume to tonnes of biomass by multiplying with wood density.

17.2 Above-Ground Biomass of Non-Trees

Non-tree biomass includes shrubs, herbs and grass biomass. Herbs and grasses are annuals, whereas shrubs are normally perennials. The focus of forest and plantation managers and greenhouse gas inventory experts is on tree biomass. However, non-tree biomass may have to be estimated for climate change mitigation projects, grassland development projects and community forestry programmes. The harvest method is used for estimating non-tree biomass in forest and non-forest land-use systems.

17.2.1 Shrub and Herb Biomass

Estimation of shrub and herb biomass may be necessary for forests and plantations. This biomass is expressed as tonnes of dry biomass production per hectare per year. Shrub and herb biomass is estimated separately, since the sample plot size as well as the form of the plants is different. Biomass for these two plant types is estimated through the harvest method.

Shrub biomass The field method for estimating the shrub biomass is described in Chapter 10. Shrub biomass includes annual as well as perennial plants. The parameter measured for the shrub biomass includes the weight of the shrub biomass in sample plots

Step 1: Record the fresh and dry weight of the shrub biomass harvested from sample plots as kg per plot

- If there are young regenerating valuable tree plants and any economically valuable perennial shrubs, harvesting such plants may not be desirable
- A few representative plants could be harvested and weighed, and the height and spread of each of these plants recorded along with the name of the species
- These data could be used for estimating the weight of plants that cannot be harvested
- Alternatively, some of the perennial or economically valuable shrub species could be ignored if they cover only a small proportion of the ground area (e.g. <10%)

Step 2: Pool all the biomass harvested from different shrub plots to obtain the total dry shrub biomass for the total area of the sample plots

Step 3: Extrapolate the sample area biomass to per hectare value (dry tonnes per hectare)

Herb biomass Biomass from herbs accounts for a small fraction of the total above-ground plant biomass of forests, plantations, agroforestry systems and grasslands. Further, herb biomass, being annual, is part of the annual carbon cycle with no implications for changes in carbon stocks. Herb biomass could be ignored, especially if tree crowns and shrubs cover the land area. If necessary, changes in carbon stocks in herb biomass could be estimated by using the method described for shrubs, based on the weight of herb biomass at two points in time. Herb biomass production (dry tonnes per hectare per year) can be calculated using the procedure given for grasses.

17.2.2 Grass Biomass Production: Above the Ground

Estimating grass biomass production (tonnes per hectare per year) is important for grassland development projects in assessing the impact of a management practice on grass production. Grass biomass is expressed as dry tonnes per hectare per year. The method of estimation is described in Chapter 10. The calculation procedure for grass and herb biomass is as follows:

Step 1: Obtain the fresh weights of grass or herbs harvested monthly from the sample plots during field studies and estimate the dry weight of samples
Step 2: Extrapolate the dry weight of grass or herb estimated on monthly basis for the sample plots area to a per-hectare value
Step 3: Tabulate or plot on a graph the grass or herb production estimated as tonnes of dry biomass per hectare for each month of the year
Step 4: The highest value among the monthly values is considered as the grass or herb productivity for that area expressed as dry tonnes per hectare per year.

17.3 Below-Ground Biomass

Methods for measuring root biomass are described in Chapter 11. The methods of excavation of roots and monoliths of soil are not practical in most situations because of high cost and the difficulty in uprooting or digging inside a forest or plantation or an agroforestry plot. Therefore, the two most common and feasible approaches are

(i) Standard root to shoot ratios (Table 17.6)
(ii) Allometric equations (Table 17.7).

(i) *Root to shoot ratio* Using root to shoot ratios to estimate root biomass involves the following steps:

Step 1: Estimate the above-ground biomass, using one of the methods described in Section 17.2, in terms of tonnes of dry biomass per hectare
Step 2: Select the appropriate root to shoot ratio from the literature. A review by Cairns et al. (1997), covering more than 160 studies from tropical, temperate and boreal forests, estimated a mean root to shoot ratio of 0.26 with a range of 0.18–0.30. Thus, for most projects, a root to shoot ratio of 0.26 could be used

Step 3: Calculate the root biomass using the data on above-ground biomass and the root to shoot ratio selected with the following formula:

$$\text{Root biomass (in dry t/ha)} = 0.26 \times \text{above-ground biomass (dry t/ha)}$$

(ii) *Allometric equations* Biomass equations have been developed to estimate root biomass using data on above-ground biomass. Examples of biomass equations are given in Table 17. 7. The method involves

- Estimating the above-ground biomass using one of the methods described in Section 17.2
- Selecting the appropriate biomass equation
- Substituting the above-ground biomass value in the equation to obtain root biomass in tonnes of dry root biomass per hectare

Table 17.6 Ratio (R) of below-ground biomass to above-ground biomass (tonnes of root dry matter: tonnes of shoot dry matter). (From IPCC 2006, compiled from various sources.)

Domain	Ecological zone	Above-ground biomass (t/ha)	R
Tropical	Tropical rain forest	–	0.37
	Tropical moist deciduous forest	<125	0.20 (0.009–0.25)
		<125	0.24 (0.22–0.33)
	Tropical dry forest	<20	0.56 (0.28–0.68)
		<20	0.28 (0.27–0.28)
	Tropical mountain systems		0.27 (0.27–0.28)
Subtropical	Subtropical humid forest	<125	0.20 (0.09–0.25)
		<125	0.24 (0.22–0.33)
	Subtropical dry forest	<20	0.56 (0.28–0.68)
		<20	0.28 (0.27–0.28)
Temperate	Temperate oceanic forest, temperate continental forest, temperate mountain systems	Conifers <50	0.40 (0.21–1.06)
		Conifers 50–150	0.29 (0.24–0.50)
		Conifers >150	0.20 (0.12–0.49)
		Eucalyptus spp. >50	0.44 (0.29–0.81)
		Eucalyptus spp. 50–150	0.28 (0.15–0.81)
		Other broad-leaved forest <75	0.46 (0.12–0.93)
		Other broad-leaved forest 75–150	0.23 (0.13–0.37)
		Other broad-leaved forest >150	0.24 (0.17–0.44)
Boreal	Boreal coniferous forest, Boreal tundra woodland, Boreal mountain systems	>75	0.39 (0.23–0.96)
		<75	0.24 (0.15–0.37)

Table 17.7 Regression equations for estimating root biomass of forests. (From Cairns et al. 1997.)

Conditions and independent variables	Equation Y = root biomass (t)	Sample size	R^2
All forests, AGB	Y = Exp[−1.085 + 0.9256*LN (AGB)]	151	0.83
All forests, AGB and age (years)	Y = Exp[−1.3267 + 0.8877*LN(AGB) + 0.1045*LN(AGE)]	109	0.84
Tropical forests, AGB	Y = Exp[−1.0587 + 0.8836*LN(AGB)]	151	0.84
Temperate forests, AGB	Y = Exp[−1.0587 + 0.8836*LN(AGB) + 0.2840]	151	0.84
Boreal forests, AGB	Y = Exp[−1.0587 + 0.8836*LN(AGB) + 0.1874]	151	0.84

LN = natural log, Exp = "e to the power of", AGB = above-ground Biomass (t); R^2 = coefficient of determination

Estimating root biomass for non-tree vegetation For non-tree vegetation such as shrubs, herbs and grasses, it is not possible to calculate below-ground biomass using above-ground biomass data. Therefore, on-site measurement using soil core method or pit method is used for the purpose, as described in Chapter 11. The root biomass for non-tree plants could be calculated as follows:

- Obtain the dry weight of root biomass by the core sample method along with the volume of the core (depth and height) of the core sampler
- Calculate the dry weight of root biomass for the volume of the core sampler calculated using the diameter and depth of the core (volume of the core = $\pi r^2 \times H$)
- Extrapolate the root biomass from the volume of the soil on per hectare basis for the depth, usually 30 cm depth (where 30 cm depth = 3,000 m^3) of soil per hectare

17.4 Calculation of Deadwood and Litter Biomass

Deadwood and litter are unlikely to be the key carbon pools for majority of land-based projects and greenhouse gas inventories. These two pools may not exist at all or exist only in insignificant quantities for grassland and agroforestry projects or even plantation forestry projects. Calculation of deadwood and litter biomass may be of significance only for forestry projects. The methods for monitoring deadwood and litter biomass are described in Chapter 12. This section describes the procedure for calculating the biomass from these two pools.

17.4.1 Deadwood

Deadwood biomass includes standing deadwood and fallen deadwood. It is important to distinguish between deadwood and litter based on the size of the wood. Further, it may not be possible to identify the species of the deadwood, especially of fallen deadwood in multispecies forests. In monoculture plantations, the dead trees could be assumed to be of the same species as the living ones that make up the plantation.

(i) *Standing deadwood*

Step 1: Obtain the values of DBH and height of the standing dead trees from field measurements. The dead trees may or may not have a crown (Chapter 12)
Step 2: Calculation of the weight of the standing deadwood

- Assume that the relation between biomass and DBH or height is the same in dead and living trees
- Adopt the biomass equation method using the DBH and, if necessary, the height using the procedure and steps given in Section 17.1.5
- Estimate the standing deadwood in dry kg or tonnes per sample plot and total the weight for all the sample plots
- Extrapolate the value from sample plot area to per hectare area

(ii) *Fallen deadwood*

- Fallen deadwood could be the whole tree, only the main stem or only large branches or a combination of these forms
- *Fallen whole tree* Adopt the method described for standing deadwood, which is same method as the biomass equation method (Section 17.1.5)
  - Estimate the dry biomass of fallen deadwood per sample plot and total the weight for all the sample plots
  - Extrapolate the value to per hectare
- *Fallen stem* Estimate the volume of the fallen stems and calculate their volume and dry weight using the steps given in Section 17.1.2
  - Estimate the volume of each stem using DBH and height
  - Estimate the weight of the stems from the wood density and volume of the stem
  - Calculate the weight of all the stems in the sample plots and aggregate it for the total sample area
  - Extrapolate the weight of total fallen stems to per hectare (dry tonnes)
- *Fallen dead branches* Biomass of fallen dead branches could be estimated as follows
  - *From the weight* Weigh all the fallen branches (excluding the litter) using a field (spring) balance, estimate the dry weight by taking a sample and drying in the oven and extrapolate the weight of the fallen branches to per hectare from the sampled area

- *From the volume* If the branches are too large for weighing with a field balance, estimate the volume using the method described for fallen stems in Section 17.1.2

(iii) *Total deadwood*

Total deadwood biomass is the sum of standing deadwood and fallen deadwood estimated using the following equation and expressed in tonnes of dry matter per hectare.

Total deadwood = standing deadwood + (fallen whole tree + fallen stem + fallen dead branches)

17.4.2 Litter

Two methods are used in estimating litter biomass, namely annual production and stock change (Chapter 12). Calculation procedures for these two methods are as follows.

(i) *Annual production* The method described in Chapter 12 will provide monthly production figures for woody and non-woody litter from the sample plots

Step 1: Obtain the monthly values of fresh and dry weight of litter from all the sample plots (in kg)

Step 2: Aggregate the monthly weights of dry litter from all the sample plots for all the 12 months

Step 3: Extrapolate the dry litter weight from sample area to per hectare value and express it as dry litter production in tonnes per hectare per year

- It is possible to estimate woody litter and non-woody litter production per hectare separately by repeating the above method for both the components of litter

(ii) *Stock change* This method of stock change requires estimation of litter stock at two points in time. The field method for measuring litter stock at a given point in time is described in Chapter 12.

Step 1: Obtain the fresh and dry weight of the litter stock in sample plots for the two periods (t_2 and t_1)

Step 2: Add the sample dry weight of all plots and extrapolate to tonnes of dry matter per hectare

Step 3: Calculate the annual litter stock change (in tonnes per hectare per year) using the following formula:

Annual litter stock change = (litter stock at period t_2 – litter stock at period t_1) / ($t_2 - t_1$)

17.5 Soil Organic Carbon

Estimation of soil carbon density (tC/ha) involves estimation of bulk density of the soil and soil organic matter content. The method for estimating these two parameters is described in Chapter 13. This section summarizes the steps involved in calculating soil carbon density

Step 1: Select the land-use category, project activity and stratum
Step 2: Conduct field and laboratory studies and estimate the bulk density and soil organic matter or carbon content (Chapter 13).

Bulk density Estimate bulk density by using tube core or clod method for undisturbed soil and the following formula:

$$\text{Bulk density (g/cc)} = \text{(weight of soil with tin} - \text{weight of empty tin)/volume of the tin}$$

or

$$\text{weight of soil clod/volume of the soil clod}$$

Soil carbon density (tC/ha) The content of organic carbon in soil estimated in percentage terms needs to be converted to tonnes per hectare using bulk density, depth of the soil and area ($10{,}000\,\text{m}^2$).

$$\text{SOC (t/ha)} = [\text{soil mass in } 0-30 \text{ cm layer} \times \text{SOC concentration (\%)}] / 100$$

$$\text{Soil mass (t/ha)} = [\text{area } (10{,}000 \text{ m}^2/\text{ha}) \times \text{depth } (0.3 \text{ m}) \times \text{bulk density } (\text{t/m}^3)]$$

17.6 Formulae and Calculations for Estimating Different Carbon Pools

The main goal of carbon inventory is to estimate carbon stocks and changes in selected land-use categories, project activities and strata. Estimating changes in carbon stock requires carbon stock values at two points in time and estimating total stocks requires estimates of changes in all the relevant carbon pools for a given land-use system. The key issues to be considered in carbon inventory calculations are as follows, which are also summarized in Table 17.8:

- The goal of carbon inventory calculation
- Type of inventory output required for the project or the programme

Table 17.8 Features of carbon (C) inventory calculations required for different projects and programmes

Project or programme	Objective of carbon inventory	Type of carbon stock estimates needed	Carbon pools to be calculated	Frequency of calculation	Method
Land-based mitigation projects	Estimation of additional carbon sequestered over baseline	Total C stock change between two points in time	All 5 pools	3–5 years	C stock-difference
Commercial roundwood production projects	Calculation of commercial merchantable biomass production	Total commercial biomass production	Largely above-ground biomass	At the end of rotation period	C stock-difference
Community forestry	Calculation of fuelwood production: trunk + twigs + branches	Total biomass production	Above-ground biomass	Annual biomass production	C gain–loss
Land reclamation or grassland development	Improvement in soil fertility	Increase in soil organic carbon	Soil organic carbon	Periodically, say once in 3–5 years	C stock-difference
Greenhouse gas inventory estimation	Calculation of carbon emissions and removals	Changes in total C stocks	All pools	Inventory year	C stock-difference or C gain–loss

- Carbon pools to be calculated and reported
- Frequency of calculation and reporting
- Approach or method to be adopted

(i) *Goals* The goal of a carbon inventory is to answer the questions relevant to the specific land-use category, vegetation type and objectives of the project or programme. Some examples of programmes and projects that require carbon inventory are

- Climate change mitigation
- Roundwood (fuelwood, timber) production
- Greenhouse gas inventory of land-use categories
- Costs and benefits analyses

(ii) *Types of inventory output* The types of carbon inventory estimates or outputs required are

- Tonnes of carbon sequestered in biomass and soil in mitigation projects
- Tonnes of carbon dioxide emissions reduced (from avoided deforestation)
- Tonnes of commercial biomass or traditional fuelwood produced
- Estimates of emissions or removals of carbon dioxide in tonnes from land-use categories

(iii) *Carbon pools* The carbon pools to be calculated and reported depend on the type of project and its objective. Projects can vary from reporting only above-ground biomass to reporting that from all the carbon pools. Some examples are

- Climate mitigation projects: all the carbon pools
- Roundwood production programmes: above-ground biomass, particularly commercial timber (main trunk)
- Community forestry: above-ground biomass, deadwood and litter
- Greenhouse gas emissions estimation: all the five carbon pools, depending on the key category analysis

(iv) *Frequency* Estimates of carbon stocks and changes are required at various phases of project cycle, namely

- Project proposal preparation
- Project implementation
- Project monitoring
- End of the project period or end of rotation period
- Annual changes in stocks

Further, carbon inventory estimates are required at different frequencies depending on the objectives of the carbon inventory projects or programmes. The reporting could be annual, periodical or only at the end of the project.

(v) *Approach or method* The approach to or methods of calculation includes the steps to be adopted based on the indicative parameters measured or observed in the field or laboratory. The calculation procedure is presented as a series of equations based on IPCC (2003, 2006) and involves estimating the changes in stocks of above-ground and below-ground biomass, deadwood, litter and soil carbon. Although five carbon pools are considered, given the resource constraints, there is a need to identify the key carbon pools for a given project or a programme or inventory.

Making a carbon inventory requires estimation of carbon stocks at two points in time or carbon gain and loss for a given year. These two methods are described using a series of equations from IPCC (2003, 2006) in Chapter 9.

In the carbon "Stock-Difference" method, carbon stock changes are calculated per year per hectare and then multiplied by the total area under each stratum to obtain the total stock changes in each pool. These values are finally aggregated. In the carbon "Gain–Loss" method, the gains and losses of different carbon pools in a given year or over a period of years are calculated.

17.6.1 Calculation of Carbon Stocks and Changes for Climate Mitigation Projects

This handbook focuses on the carbon "Stock–Difference" method for evaluating the impact of a mitigation project on carbon stocks, under which only the relevant carbon pools are included in the baseline and mitigation scenarios. The calculation procedure involves the following steps:

Step 1: Estimate the carbon stock change in baseline scenario for the period considered:

$$\Delta \mathrm{CB} = (\mathrm{CB}_{t_2} - \mathrm{CB}_{t_1})$$

where:
ΔCB = change in carbon stock in baseline scenario in tC during the period $t_2 - t_1$
CB_{t_2} = total carbon stock in the baseline scenario during the year t_2 (e.g. at the end of the 5th year)
CB_{t_1} = total carbon stock in the baseline scenario during the year t_1 (year zero or the year the project began)
If a stable baseline scenario is assumed, carbon stock is estimated at the beginning of the project t_1 which is t_0.

$$\Delta \mathrm{CM} = (\mathrm{CM}_{t_2} - \mathrm{CM}_{t_1})$$

where:
ΔCM = change in carbon stock in mitigation scenario in tC between the period $t_2 - t_1$
CM_{t_2} = total carbon stock in the mitigation scenario during the year t_2 (e.g. at the end of the 5th year)

CM_{t_1} = total carbon stock in the mitigation scenario during the year t_1 (year zero or the year the project began)

Step 2: Estimate the carbon stock change in project scenario for the period considered

Step 3: Estimate the additional or incremental carbon stock change in mitigation scenario over the baseline scenario

$$C_{AD} = (\Delta CM - \Delta CB)$$

where:

C_{AD} = additional or incremental carbon stock change in tC during the period $t_2 - t_1$ (where t_2 is the year for which the calculations are made, for example, end of 5th year)

ΔCM = change in carbon stock in mitigation scenario in tC during the period $t_2 - t_1$

ΔCB = change in carbon stock in baseline scenario in tC during the period $t_2 - t_1$.

17.6.2 National Greenhouse Gas Inventories

IPCC (2003, 2006) provides detailed guidelines for estimating carbon emissions and removals from land-use categories, which are summarized in Chapter 9. Carbon inventory can be prepared using the two methods presented in Chapter 16, namely carbon "Stock-Difference" and carbon "Gain–Loss".

17.6.3 Calculation of Biomass Stocks and Changes for Roundwood Production Projects

Roundwood production projects require monitoring and estimation of biomass stocks at different periods and particularly at the end of the rotation. The carbon stocks could be calculated using the "Stock-Difference" method, as follows:

$$RW_p = (RW_{AGB_{t_2}} - RW_{AGB_{t_1}})$$

where:

RWp = total roundwood production (above-ground biomass in tonnes)

RW_{AGBt_2} = roundwood stock at year t_2 (above-ground biomass in tonnes)

RW_{AGBt_1} = roundwood stock at year t_1 (above-ground biomass in tonnes)

Commercial industrialwood plantations require estimates only of above-ground biomass pool, and those too are limited to only merchantable tree trunk (excluding tree twigs and branches) – deadwood and litter pools are not required.

Fuelwood and community forestry plantations require estimates of above-ground biomass pool including tree trunks, twigs and branches; deadwood and woody litter could also be included if necessary since these pools also supply traditional fuelwood.

17.6.4 Carbon Inventory for Agroforestry, Shelterbelt, Grassland Management and Soil Conservation Activities

Carbon inventory is required for agroforestry, shelterbelt, grassland management and soil conservation projects and activities. The methods for estimating carbon from biomass and soil are similar to those described in earlier sections. This section presents the key features and the carbon pools to be included.

(i) *Agroforestry* Agroforestry is often a component of village ecosystems involving a large number of farms. Agroforestry projects aim to enhance: (i) the density and diversity of trees and carbon stock in soil and vegetation, (ii) flow of tree-based products and incomes and (iii) crop productivity. Crop production remains the dominant activity, with rows of trees in the middle of a crop or along the bunds or boundaries of fields.

Carbon pools Above-ground tree biomass is the most important carbon pool. Soil organic carbon needs to be measured only if the agroforestry activity involves planting a large number of trees or rows of trees spaced densely, although it is difficult to suggest a specific number.

(ii) *Shelterbelt* Shelterbelts comprise rows of trees at the boundary of a village or a block of farms to check wind erosion, halt desertification, enhance carbon stocks, possibly increase biomass (fuelwood and non-wood tree products) and ultimately increase crop productivity.

Carbon pools Above-ground tree biomass is the only critical carbon pool to be measured or monitored. Below-ground biomass can be estimated using root to shoot ratio. Because the density of trees per hectare is low, other carbon pools may not be relevant.

(iii) *Soil conservation practices* Watershed protection including soil conservation is one of the critical objectives of many land development projects. Watershed protection is achieved by soil conservation practices such as contour bunding, gully plugging, and check dams. Soil conservation measures also increase soil organic matter and productivity of crops or grass cover.

Carbon pools The only carbon pool that will be impacted is soil carbon.

(iv) *Grassland management practices* Grassland or pasture land or rangeland management practices, including soil and water conservation, planting grasses, regulation of grazing or grass harvest and fire control, could lead to increased grass productivity and increased soil carbon density.

Carbon pools The most important carbon pool to be measured or monitored is soil carbon, which will be impacted most by grassland management practices.

17.7 Data and Sources of Data

Estimation of carbon stock changes or emissions and removals of CO_2 at a project or national level for a given year or over a period of years requires data on the area under a selected land-use category or project activity and estimates of carbon stock or rate of change per unit area. The reliability of estimates depends on the quality of the data on area as well as factors that affect carbon stocks and changes in the stocks. This handbook focuses on estimation of stocks and rates of changes of different carbon pools. The values for the stocks and rates of change in different carbon pools are highly variable for different forest, plantation, grassland and agroforestry systems, based on rainfall, soil type, topography, species and management systems. Estimates of carbon stocks and changes for land-use systems are characterized by high variation and possibly uncertainty. The values of carbon stocks and rates of change for different carbon pools should be ideally measured and estimated for different locations. However, it may not always be feasible to measure the different carbon pools at the project level or land-category level and particularly at the national level. Thus, the values for different carbon pools may have to be obtained from literature, databases and measurements carried out at other locations with similar biophysical conditions.

Carbon inventories at project as well as national level are often carried out using default or literature-based values. Default values are factors or data obtained from experiments and from field and laboratory measurements or from studies of stocks and growth rates of different carbon pools using statistical procedures and published in databases, reports, books and journal articles. The IPCC guidelines (IPCC 1996, 2003, 2006) provide Tier 1 methods for estimating CO_2 emissions and removals, which are largely based on default values. Default values from literature, models and experiments are indispensable to carbon inventories, particularly in the following situations:

- The project development phase requires default values for all the carbon pools to project gains in carbon stocks or roundwood production.
- The project monitoring phase requires default values for some carbon pools or conversion factors such as biomass conversion and expansion factors and root to shoot ratios since these values are difficult to measure in all the projects or locations.
- A national greenhouse gas inventory involves estimation of CO_2 emissions and removals for all land-use categories and all the relevant carbon pools, and it is very rarely that the required values of carbon stock changes for all land-use categories are measured.

- Economic analyses of roundwood production programmes often measure a few parameters and use models and default data to estimate costs and benefits from wood production.

The main approach that any project developer or greenhouse gas inventory expert should take is to generate location-specific values for different carbon pools and use default values only when it is not feasible or too expensive to locally estimate the values for carbon stocks and changes. However, most often default values or conversion factors are used for one or more carbon pools by project developers, monitoring experts and GHG inventory compilers. Any carbon inventory expert must be familiar with different sources of carbon emissions and removals and default values of the rates of stock change. The inventory expert may have to scan all the default data, evaluate or validate them based on expert judgment and use the values. This section discusses the input data needed, different sources of default data and their use in carbon inventory.

17.7.1 Data Needs for a Carbon Inventory

- Data needs for a carbon inventory first include activity data on area under different project activities, land-use categories and subcategories at the national level. Secondly, estimates of stock and rates of change in stocks are also needed for different carbon pools. The key carbon inventory parameters required, related to the stocks and rates of change of different carbon pools and a few factors or constants.
- Growth rates of biomass and soil carbon pools.
- Factors such as density of wood, bulk density of soil and biomass conversion and expansion factor.
- Allometric equations for estimating the volume or biomass for above-ground and below-ground biomass.
- Grass production.

Examples of types of projects and programmes requiring default values are as follows:

- *Forestry mitigation projects* afforestation, reforestation, avoided deforestation, bioenergy plantations and agroforestry
  - Above-ground and below-ground biomass, deadwood and litter and soil organic carbon: stock and growth rates and allometric equations
  - *Grassland management projects* soil organic carbon and below-ground biomass
- *Timber or roundwood production projects* above-ground commercial timber volume or total above-ground tree volume
- *Bioenergy plantation (not linked to mitigation)* above-ground biomass production

- *Greenhouse gas inventory* above-ground and below-ground biomass, deadwood and litter and soil organic carbon: stock and growth rates, and allometric equations
    - Forest land, grassland, cropland, settlement land-use categories
    - Land remaining in the same category and land converted to another category

17.7.2 Sources of Data

The values of carbon inventory parameters can be obtained from secondary sources and primary or location-specific measurements. This chapter focuses on secondary sources, which include published reports on forest inventories, unpublished studies and reports, national and international databases, web sites, national communications, GHG inventory reports and other literature.

(i) *Forest inventories* Studies on forest inventories conducted in many countries are an important source of data on biomass-related parameters. Normally forest inventories are a good source of data on above-ground biomass, in particular, the merchantable volume production. Such reports should be explored first, although they may suffer from the following limitations:

- Repeated measurements may not be available for the same plots to estimate the average growth rates
- Below-ground biomass, litter, deadwood and soil organic carbon may not be measured
- Non-forest land-use categories may not be included
- Biomass data may not be recorded for all the tree species, particularly non-commercial tree species

(ii) *IPCC inventory guidelines* The *Revised 1996 IPCC Guidelines*, *Good Practice Guidance* (IPCC 2003) and *2006 IPCC Guidelines* provide default values for most of the emission and sequestration factors for different land-use categories required for a carbon inventory, which is part of the national greenhouse gas inventory. These default values have gone through an intensive review process as well as referenced to the original source of the data. The data in the latest IPCC 2006 guidelines are organized according to the following method of stratification, wherever permitted by the original source of data.

- Climate domains: tropical, subtropical, temperate, boreal and polar
- Climate regions within each climate domain
    - Tropical: tropical wet (tropical rain forest), tropical moist (tropical moist deciduous forest), tropical dry (tropical dry forest, tropical shrubland and tropical desert), tropical montane (tropical mountain systems)
    - Subtropical: warm temperate moist, warm temperate dry
    - Temperate: cool temperate moist, cool temperate dry

- ◦ Boreal: boreal moist, boreal dry

Further, default values are also provided for different carbon pools by continent, forest type and tree species. Default values are provided for the following emission/removal factors and conversion factors in the IPCC 2006 guidelines:

- Carbon fraction of above-ground forest biomass
- Ratio of below-ground to above-ground biomass
- Biomass expansion factor
- Wood density
- Above-ground biomass in forests and plantations
- Mean annual increment in forest and plantation species
- Litter and deadwood carbon stocks
- Soil organic carbon stocks, soil carbon stock change factors for land-use categories
- *Agroforestry systems* above-ground biomass for perennial crops, biomass accumulation rates
- *Grassland* stock change factor for land conversion, above-ground biomass stocks

The default values given in the IPCC inventory reports (IPCC 1996, 2003, 2006) can be used not only for carbon inventory for the national greenhouse gas inventory, but also for mitigation projects in the land-use sectors as well as for roundwood production projects. However, some of the following shortcomings of the IPCC default values limit their application.

- Default values refer largely to an ecological zone (e.g. tropical dry forest or tropical rain forest) and at the continent level, except for wood density and mean annual increment
- Default values are not provided at national or regional levels for different species, age of the stands, density of plantations and other management practices

(iii) *Emission factor database (EFDB)* A database of emission factors is being initiated and coordinated by the IPCC greenhouse gas inventory programme mainly to assist national greenhouse gas inventory experts. The EFDB is meant to provide well-documented emission and sequestration factors as well as other related parameters for making GHG inventories and to establish a communication platform for distributing and commenting on new research and measurements of data relevant to greenhouse gas inventories. The criteria for inclusion of data in the database are as follows (IPCC 2006):

- *Robustness* The value should be unlikely to change, within the accepted uncertainty of the methodology, if the original measurement programme or modelling activity is repeated.
- *Applicability* An emission or sequestration factor can only be applicable if the biophysical or management conditions from which the values originate are known to enable the user to decide where and when the data can be applied.
- *Documentation* Access to information in the original technical reference is provided to evaluate the robustness and applicability of the data.

The EFDB is an online database, which is being continually updated with data reviewed by a panel of experts. It is a menu-driven, user-friendly database. The steps involved in obtaining the required values from this database are listed below:

Step 1: Selection of the sector such as land-use change and forestry
Step 2: Selection of land-use category such as forest land and grassland
Step 3: Selection of a carbon pool
Step 4: Selection of a filter giving such conditions as the region, species and rainfall zone
Step 5: Downloading the default values for the parameter

The EFDB mostly contains values from various IPCC reports and some other well-known sources of data. The database is expected to grow into a full-fledged database to assist greenhouse gas inventory experts as well as land-based project developers. Experts and researchers all over the world are invited to contribute to the database with their data, which will be reviewed by an editorial board and stored in the database.

(iv) *FAO database*: The FAO is a source of extensive data on forests, agricultural crops, agroforestry systems and grassland ecosystems at global and national levels. Data are also provided by forest ecological zones and by crops. The FAO is a main source of data on:

- Area under different crops, food production and crop productivity at national level
- Land-use pattern at national and global level, by use
- Area under forests and wooded land by forest type
- Deforestation and afforestation rates
- Roundwood (industrialwood, sawnwood and fuelwood) production, consumption, import and exports

The parameters related to the carbon inventory are provided by forest ecological zones and tree plantations at national level. The carbon inventory parameters and the other features available from FAO sources are as follows:

- Carbon inventory parameters
    - Growing stock
    - Mean annual growth rates
- Level at which inventory parameters are available
    - National level
    - Forest ecological zone level
    - Plantation species
- Carbon pools for which data are available
    - Above-ground biomass

Thus, the FAO is an important source of activity data and emission and sequestration factors for land-use sectors. The FAO data can be accessed at www.fao.org.

(v) *National GHG inventory reports* All the Annex-I or industrialized countries have been submitting GHG inventories annually and majority of non-annex-I or developing countries have submitted their GHG inventory reports. Many countries have conducted field and laboratories studies, made measurements and used the values in their inventories. Carbon inventory experts could select the countries with identical or comparable biophysical conditions as well as land-use categories or forest or plantation or grassland systems and use the relevant default values. The reports can be downloaded from www.unfccc.int.

(vi) *Published books, reports and journal articles* A number of books and reports provide information and data on carbon inventory parameters. In addition, many scientific journals regularly publish articles providing data on various parameters (Table 17.9). Some examples of literature which provide information on carbon inventory parameters are listed here.

- Cannell (1982) *World Forest Biomass and Primary Production Data*, Academic Press
  - Values for above-ground and below-ground biomass and dead organic matter for a large number of forests and plantations
- Cairns et al. (1997) Root biomass allocation in the World's upland forests, *Oecologia*
  - Values for above-ground and below-ground biomass and root to shoot ratios
- Mokany et al. (2006) Critical analysis of root to shoot ratios in terrestrial biomes, *Global Change Biology*
- Pearson et al. (2005b)
  - Biomass estimation equations
- Brown (1997)
  - Biomass estimation equations
- Reyes et al. (1992) Wood densities of tropical tree species, USDA
  - Wood densities of tropical tree species
- Ugalde and Perez (2001) Mean annual volume increment of selected industrial forest plantation species, FAO
  - Above-ground biomass growth rates of forest plantation species
- Forest Survey of India (1996)
  - Biomass estimation equations

Table 17.9 Global data sources, parameters included and applications

Source	Factors/parameters/ equations	Features	Applications
FAO www.fao.org	- Area; forest, agriculture, etc. - Deforestation/ afforestation - Biomass stocks	- Periodic estimates of area and changes - Forest type	- Roundwood production and mitigation projects - GHG inventory estimation
IPCC 2006 Guidelines www.ipcc.ch	- Biomass and SOC stocks - Growth rates of C pools - Biomass conversion and expansion factors - Wood density	- FAO vegetation zone - Periodically produced - Country-level data	- GHG inventory estimation - Mitigation projects
EFDB www.ipcc-nggip.iges.or.jp	- Biomass and SOC stocks - Growth rates of C pools - Biomass conversion and expansion factors - Wood density	- Data evaluated based on criteria by an expert body	- GHG inventory estimation - Mitigation projects
National GHG inventory reports http://unfccc.int/national_reports/items/1408.php	- Biomass and SOC stocks - Growth rates of C pools - Biomass conversion and expansion factors - Wood density	- Data could provide values from national sources	- National GHG inventory

- Murali and Bhat (2005)
  - Biomass estimation equations for tropical deciduous and evergreen forests

(vii) *Studies in the region* Mitigation project developers and GHG inventory experts could examine results of studies conducted in the region or in the vicinity of the project or land-use category. Studies conducted on forests, plantations, grasslands and agroforestry systems in the region of interest may have monitored various carbon inventory parameters such as above-ground biomass stocks and soil organic carbon content. If the study location has identical or comparable rainfall, soil, topography, species mix, etc., the data generated would be highly suitable for mitigation or roundwood production projects and even the GHG inventory programme. Such data obtained from studies conducted in the region could be more applicable than the global or even national databases. However, it is important to verify the methods of sampling, measurement and calculation adopted by such studies.

17.7.3 *Criteria for Evaluation and Selection of Parameters*

The default values provided in many databases such as IPCC, FAO and EFDB are often based on a single or a few studies for a given forest or plantation type. The values normally provide global or national averages for a given parameter for a given land-use category or forest or plantation type. Often, multiple sources provide the values for a given parameter such as above-ground biomass growth rate or soil carbon density. For example, above-ground biomass growth rate for eucalyptus plantation is available from local, regional, national and international sources. Further, several studies may be available for a given region. It is important to ensure that extremely high or low values for a given carbon inventory parameter are not used, even if they are from the region. Thus, the carbon inventory expert may have to compare different values available, validate them and make expert judgment before selecting a value for calculation of carbon stocks or changes. It is important for the carbon inventory expert to stratify the project location or land-use category and to select appropriate default values. Further, even if a parameter is actually measured for a given location, it is desirable to compare the results with the values reported in literature to validate the measured values.

Values for the carbon inventory parameters are required at as disaggregated a level as possible. Given the high variability due to various biophysical factors, the location specificity of these values is important. However, it is also important to state here that for many parameters the average values may indeed be adequate for different situations. For example, a review of 160 studies (Cairns et al. 1997) on root to shoot ratios showed that the values were within a small range of 0.18–0.30, and thus using a mean value of 0.26 may be adequate for most studies. The criteria for selecting the values from published or unpublished sources could be based on biophysical and management systems. Some potential stratification criteria are given in Table 17.10.

Table 17.10 Criteria for data selection for different types of forests or plantations

Climate domain	Rainfall zone (annual rainfall, cm)	Forest, plantation type or dominant species	Age of stand (years)	Silviculture or management system
- Tropical	- Humid (>200)	- Tropical wet	- <5	- Density of plantations/forests
- Subtropical	- Subhumid (100–200)	- Tropical dry	- 5–10	
- Temperate		- Warm temperate dry	- 11–20	- Rates of fertilizer application
- Boreal	- Semi-arid (50–100)	- Warm temperate wet	- 21–100	
			- >100	- Thinning
	- Arid (<50)	- Eucalyptus		- Irrigation
		- Pine		
		- Tectona		
		- Acacia		

17.7.4 Steps in Selection and Use of Carbon Inventory Parameter Values

Multiple sources of information may be available for a given carbon pool or a parameter required for a carbon inventory. It is necessary to select the most appropriate value for a given parameter. Even when only one source or value is available, the inventory expert has to make a judgment on its applicability. The following steps could be adopted for selection and application of default values or literature-based values for carbon inventory estimation:

Step 1: Select the land-use category for GHG inventory or project activity for mitigation or roundwood production programme

Step 2: Select the stratum or the land subcategory or plantation species or age of the stand

Step 3: Select the carbon inventory parameter such as above-ground biomass stock or growth rate, soil organic carbon density or wood density

Step 4: Conduct key category analysis to identify if the stratum and the carbon pool is a key category

Step 5: Define the biophysical and management system of the stratum for which carbon inventory parameter values are required

- Rainfall zone
- Soil type
- Forest or plantation type, dominant species
- Age and density of the stand
- Fertilizer application, thinning or irrigation

Step 6: Search all the potential sources of information for the selected carbon pool or parameter

- National, regional and global databases
- National forest inventory reports
- Published papers, books and reports
- National greenhouse gas inventory reports
- Project reports from the region

Step 7: Tabulate all the values available for the selected carbon pool or parameter

Step 8: Calculate the mean and standard deviation for the parameter

- Identify the outliers or extreme values, which are beyond two standard deviations, for example

Step 9: Adopt quality control procedures and check for the following in the original studies

- Methods adopted for measurement and calculation in the original studies
- Documentation and assumptions used in the studies

- Biophysical factors of the location of the original study such as rainfall, soil type, age or density of stand
- All conversions and units used
- Error and uncertainty estimates if provided

Step 10: Select the value for the carbon inventory, which could be the mean value or even a single most appropriate value from among the list of sources.

17.8 Conclusions

Carbon inventory is required for most land-based projects aimed at carbon mitigation, roundwood production, community forestry, grassland and degraded land reclamation, agroforestry and national GHG inventory programmes. Earlier chapters presented methods for measurement and monitoring of different indicator parameters for different carbon pools, such as DBH, height, bulk density, wood density and soil organic matter concentration. This chapter presents the approach and methods for calculating carbon stocks of different pools as well as total carbon stocks per hectare and total area. Identifying suitable biomass estimation equations or developing location- and species-specific equations is critical to calculating above-ground and below-ground biomass for tree-based land-use systems. Deadwood and litter pools, which are unlikely to be key pools, could be calculated by estimating the stocks at two points in time and calculating the difference. Soil organic carbon is calculated using soil carbon concentration and bulk density values for a given soil depth. Calculation procedures for most projects such as grassland development and degraded land reclamation, agroforestry and shelterbelt programmes require estimates of differences between the stocks of relevant carbon pools at two points in time, and estimate annual stock changes by dividing the difference between the two periods by the number of years. The frequency of calculation and reporting of carbon stock change could be different from that of the measurements. Calculation and reporting are most often required annually, at the end of rotation or at the end of the project period and also at fixed intervals such as once in 5 years. Thus, estimation of annual changes in stocks of carbon facilitates calculation for any frequency.

The quality and reliability of carbon inventory estimates depend on the quality and suitability of various parameters used in the inventory. Most mitigation projects, roundwood production programmes and national GHG inventory processes require the use of published or unpublished values of various inventory parameters. The ideal approach to follow would be to generate location-specific values for the parameters using the methods described in this handbook. However, it may not be feasible or it may be too expensive to generate all the parameters required for a project or national GHG inventory. A number of sources, particularly the global databases, are available and are being used. It may not be necessary to generate some parameters locally, since they may not vary much from the published average values such as root to shoot ratio, wood density and biomass conversion and expansion factors. Even when values for a given parameter are generated locally, it is desirable to compare them with published values. The inventory experts should make an expert judgment before using any published or unpublished values by studying and comparing the values, the methods used and estimates of error or uncertainty, if any.

Inventory experts should explore all sources of information and data available for a carbon pool or a parameter from global to national to local sources. The importance of databases is being realized and several global and regional databases are being developed and made accessible to carbon inventory experts around the world. The quality of the carbon inventory is likely to improve in years to come with the availability of high-quality global, regional and national databases incorporating carbon inventory parameters as well as other relevant factors and data. It is important to recognize that the inventory parameters and factors (such as biomass stocks and growth rates, wood density and soil organic matter density) required for multiple purposes such as making a carbon inventory are also required for other forest conservation and development programmes, commercial or subsistence timber or fuelwood production and grassland reclamation. Thus, researchers should generate carbon inventory parameters for different regions and for multiple applications.

Chapter 18
Uncertainty Estimation, Quality Assurance and Quality Control

Carbon stocks and changes should be neither overestimated nor underestimated as far as can be judged (IPCC 2000). The uncertainty is normally high in biological and land-use sectors given the large variation in factors contributing to carbon stocks and changes. Uncertainty in the estimates that make up a carbon inventory is considered as a barrier in land-use sector, particularly in forest sector, to mitigating climate change. The uncertainty in estimated carbon stock and changes in land-use sectors is often estimated to be 25–70% of the actual values, which has to be considered high. As a consequence, assessing the reliability and accuracy of the estimated carbon stocks and changes becomes a critical requirement, and the goal of any carbon inventory programme should be to minimize such uncertainty. It is always desirable that reported estimates of carbon stocks and changes are accompanied by estimates of uncertainty: most often, this is not so because of the complexities involved in estimating uncertainties; also, not all sources of uncertainty can be quantified. The uncertainties are in many cases so high that project managers or inventory compilers hesitate to estimate and report them. A broad approach to reducing uncertainty involves the following steps:

- Identify the types and sources of uncertainty, such as lack of data, measurement error, sampling error and model limitations
- Quantify, aggregate and estimate the extent of uncertainty by source
- Increase the sample size, adopt correct statistical sampling methods, improve the representativeness of samples, reduce or avoid measurement errors and improve the model

The uncertainty analysis should help prioritize the efforts to reduce uncertainty in the estimates of carbon stocks and changes in carbon mitigation projects, roundwood production and national greenhouse gas inventory programmes. All these activities involve repeated multiperiod or multiyear and long-term measurement, monitoring and estimation of carbon stocks and changes. IPCC (2003, 2006) provides approaches to and methods for identification, estimation and reduction of uncertainty in greenhouse gas (GHG) inventory programmes. These approaches and methods could be adopted in estimating uncertainty in estimates of carbon stocks and changes in land-based climate change mitigation projects and roundwood production programmes.

18.1 Causes of Uncertainty

Estimates of carbon stocks and changes made using field methods, laboratory techniques and models are likely to differ from the true values on the ground because of several factors. Some causes of uncertainty, such as sampling error, may generate well-defined and quantified estimates of the range of potential uncertainty; other causes such as bias in locating a sample plot or in choosing a biomass estimation equation may be much more difficult to identify and quantify (Rypdal and Winiwarter 2001). Uncertainty could arise from various parameters required for estimating carbon stocks and changes including

- Above-ground and below-ground biomass stocks and growth rates, litter, deadwood stocks and soil carbon density
- Biomass expansion factors for converting commercial biomass to total tree biomass
- Wood density
- Model or equation coefficients

Attempts must be made to account for and estimate all the causes of uncertainty, both quantifiable and others. Some of the causes of uncertainty are listed below (IPCC 2006):

(i) *Lack of completeness* Incomplete observations or recording is common wherever field measurements are involved. Some of the observations, such as diameter at breast height (DBH) and height, may be missing, incomplete, not recorded or not entered in the database. Data on some carbon pools, such as litter or below-ground biomass, may be missing or may not be recorded.

(ii) *Lack of data* Data on some of the parameters required may not be available. For example, data on height of trees may not be collected because of the difficulty in measuring the height of tall trees in a mature forest or plantation although a suitable biomass estimation equation may require DBH and height values. Similarly, litter data may not be available and further, no default values may be available for the vegetation types selected. Absence of such data may lead to incomplete estimation of carbon stocks or changes.

(iii) *Lack of representativeness* Lack of representativeness as a source of uncertainty is associated with the lack of complete correspondence between conditions associated with the collected data and those with vegetation on the ground. Further, the sample plots selected for a given forest or grassland type may not fully represent the variation in the field due to

- Differences in the age of the forest stands
- Variation in tree density
- Changes in species mix
- Variation in soil type and topography
- Management practices

(iv) *Statistical sampling error* Sampling error as a source of uncertainty is associated with data obtained from a sample of finite size determined not by the variance of the population – as it should be – but by factors such as limitations of resource and time. Errors may also be due to the location of the sampling unit.

(v) *Measurement error* Errors in measuring may be random or systematic and result from defective instruments or techniques of measurement or may creep in during recording and transmission of information.

(vi) *Laboratory estimation error* Errors in laboratory estimations could occur because of impurities in reagents and poor calibration of instruments.

18.2 Estimation of Uncertainty

Estimated carbon stock changes or CO_2 emissions and removals from land-use activities have uncertainties associated with area or other activity data, biomass growth rates, expansion factors and other coefficients. This section describes how to estimate and combine these uncertainties at the carbon pool or land-use category level and also how to estimate uncertainty in the inventory as a whole. It assumes that estimates of uncertainties in input data are available as default values, based on expert judgement or sound statistical sampling.

Uncertainties should be reported as a confidence interval giving the range within which the underlying value of an uncertain quantity is thought to lie with a specified probability. The IPCC Guidelines (2003) suggest the use of a 95% confidence interval, which is the interval that has a 95% probability of containing the unknown true value. This may also be expressed as a percentage uncertainty, defined as half the confidence interval width divided by the estimated value of the quantity. Percentage uncertainty is applicable when either the underlying probability density function is known or when a sampling scheme or expert judgement is used. Probability density function (PDF) is a function that represents the probability distribution in terms of integrals and is represented by a curve such that the area under the curve between two numbers is the probability that the random variable will be between those two numbers. The most common PDF is the normal distribution. Uncertainty is estimated for parameters such as biomass and soil carbon, wood density and carbon fraction of dry matter, which can be assessed from the standard deviation of measured sample values, and is estimated using the following equation

$$\frac{Us\ (\%) = \tfrac{1}{2}\ (95\%\ \text{confidence interval width}) \times 100}{\mu}$$

where:

Us = percentage uncertainty in the estimate of mean parameter value
μ = sample mean value of the parameter

IPCC Good Practice Guidance (IPCC 2003) suggests two methods for estimating and aggregating uncertainty. Uncertainty is also often estimated based on expert judgement.

(i) Simple error propagation
(ii) Monte Carlo simulations

18.2.1 Simple Error Propagation

The error propagation equation yields two convenient rules for combining uncorrelated uncertainties under addition and multiplication. First, where uncertain emission factors and activity data values (quantities) are to be combined by multiplication, the standard deviation of the sum will be the square root of the sum of squares of standard deviations of the quantities that are added. The standard deviations are expressed as coefficients of variation, which are the ratios of the standard deviations to the appropriate mean values and give the percentage uncertainty associated with each of the parameters. This rule is approximate for all random variables. Under typical circumstances, the rule is reasonably accurate as long as the coefficient of variation is less than approximately 0.3 for each parameter. A simple equation can then be derived for the uncertainty of the product, expressed in percentage terms.

$$U_{\text{total}} = \sqrt{U_1^2 + U_2^2 + U_3^2 \ldots U_n^2}$$

where:

U_{total} = percentage uncertainty in the product of the quantities (half the 95% confidence interval divided by the total and expressed as a percentage)
$U_{1\ldots n}$ = percentage uncertainty associated with each of the parameters ($1\ldots n$)

Secondly, where uncertain quantities are to be combined by addition or subtraction, the standard deviation of the sum will be the square root of the sum of squares of standard deviations of the quantities that are added with the standard deviations, all expressed in absolute terms (this rule is exact for uncorrelated variables). Using this interpretation, a simple equation can be derived for the uncertainty of the sum, expressed in percentage terms, and obtained using the following simple error propagation equation where uncertain quantities are combined by addition or subtraction for deriving the overall uncertainty of the project:

$$\frac{U_{\text{E}} = \sqrt{(U_1 * E_1)^2 + (U_2 * E_2)^2 + (U_{\text{n}} * E_{\text{n}})^2}}{|E_1 + E_2 + \ldots E_{\text{n}}|}$$

where:

U_{E} = percentage uncertainty of the sum
U_{i} = percentage uncertainty associated with source/sink i
E_i = emission/removal estimate for source/sink i

This simple error propagation equation assumes that there is no significant correlation among various parameters and estimates and that the uncertainties are relatively small. However, the equation can be used to get approximate estimates even when uncertainties are relatively large. The Monte Carlo simulation method can be used to overcome the limitations related to correlation between different parameters, such as above-ground biomass stock and estimates of below-ground biomass carbon stock at different periods.

18.2.2 Monte Carlo Simulations

The Monte Carlo method is suitable for forestry and other land-based projects, particularly since it helps to overcome the limitation due to lack of independence between various uncertainty values involved in assessing uncertainty at the project level. A general description of the Monte Carlo method is given in Fishman (1995), and statistical packages are available that include Monte Carlo algorithms. Winiwarter and Rypdal (2001) and Eggleston et al. (1998) provide examples of application of Monte Carlo analysis, which is designed to select random values of estimated parameters from PDFs and then calculate the corresponding change in carbon stocks. This procedure is repeated several times to obtain a mean value and the range of uncertainty. Data used for uncertainty estimation can be derived from field study or from expert judgement and need to be synthesized in such a way as to produce PDFs.

The approach to estimating uncertainties, based on *Good Practice Guidance* (IPCC 2003), consists of five steps; the first two require some effort from the user but the rest are carried out automatically by the software package.

Step 1: *Specify uncertainties in the input variables* Specifying uncertainties in the input variables includes input parameters, their associated means and PDFs and any correlations.

Input variables To provide uncertainties for the key parameters, it is desirable to independently assess the uncertainty associated with the data used in estimating carbon stocks. Ranges of uncertainty estimates for many of the input parameters required for estimating carbon stocks and changes are given in IPCC (2003, 2006) and can also be obtained from the literature, comparison with other project locations with similar conditions and even expert judgement. Uncertainty estimates can also be obtained for models (IPCC 2003).

Probability distribution function Simulating PDF requires the analyst to specify probability distributions that reasonably represents each model input for which the uncertainty is to be quantified. The function can be obtained by a variety of means including statistical analysis of data and expert judgement (IPCC 2000a). The critical step is to develop the distribution for input variables.

Step 2: *Set up software package* The emission inventory calculation, the PDFs, and the correlation values should be set up in the Monte Carlo package. The software performs the subsequent steps. In some cases, the carbon inventory agency may decide to set up its own programme to run a Monte Carlo simulation, which can be done using statistical software packages.

Step 3: *Select input values* Normally input values are the estimates applied in the calculation. This step is the start of the iterations. For each item in the input data, a number is randomly selected from the PDF of that variable.

Step 4: *Estimate carbon stocks* The variables selected in Step 3 are used to estimate carbon stocks for the base year and the current year (that is the beginning and the end of the carbon inventory period, e.g. year t_1 and year t_{10}) based on input values.

Step 5: *Iterate and monitor results* The calculated total from Step 4 is stored, and the process repeated from Step 3. The mean of the totals stored gives an estimate of the carbon stock, and the variability represents uncertainty. Many repetitions are needed for this type of analysis. The duration over which the iterations are conducted can be determined in two ways: the iterations continue until either a stipulated number – e.g. 10,000 – of runs is completed or the mean reaches a relatively stable point.

18.3 Uncertainty Analysis

Uncertainty analysis involves the following steps, which are illustrated in IPCC (2003):

Step 1: Estimate emissions or removals related to each activity – forestland remaining forest, grassland converted to forest, cropland converted to forest and so on

Step 2: Assess uncertainties related to each activity – forest land remaining forest land, grassland converted to forest land and so on

Step 3: Assess the total uncertainties from the LULUCF sector – forest land remaining forest land, grassland converted to forestland and so on

Step 4: Combine LULUCF uncertainties with other source categories such as energy and agriculture sectors.

18.4 Reducing Uncertainty

According to IPCC (2006), uncertainties should be reduced as far as practicable during the process of carbon inventory and it is particularly important to ensure that the model and the data collected are a fair representation of the reality on the ground. Efforts to reduce uncertainty should focus on input parameters and models critical for estimation or projection of carbon stocks and changes. Attempts to reduce uncertainty should be preceded by an analysis of the sources of uncertainty and the extent of uncertainty associated with key input parameters and the methods used to generate them. Some of the potential options to reduce uncertainty are as follows (IPCC 2006):

(i) *Improving conceptualization* It is necessary to conceptualize carbon inventory process to ensure consideration of all aspects relevant to reducing uncertainty, such as

- Stratification of land-use category.
- Sample size and location of sampling units.
- Method of measurement of parameters in the field and recording of data.
- Selection of a model for analysis.
- Input parameters for the model or steps in the calculation.

(ii) *Improving the models* Selection of an appropriate model for the specific project type and location, improvements in model structure and parameterization would lead to a better understanding of the model characterization and contribute to reducing the uncertainties.

(iii) *Improving representativeness* It is very important to ensure that the sample size and location provide a good representation of the carbon status in the field. Representativeness of the sample and the parameters required for estimating carbon stocks and changes are related to stratification, sample size, sample location and parameters selected for measurement. Increasing the sample size normally improves representativeness, though not always.

(iv) *Precise methods of measurement* Errors in measurement can be reduced by using more precise measurement methods and ensuring that the instruments are correctly calibrated.

(v) *Complete measurement* It is important to ensure that field studies are conducted for all the sampling units and that the identified parameters are measured in all the plots and for all the individuals, such as trees and shrubs.

(vi) *Correct recording and transmission of data* Data recorded in the field should be correctly coded and entered into a computerized database ensuring correct conversions (kilograms to tonnes, inches to centimetres or metres, etc.) and uniform units of area, weight and height.

(vii) *Quality control and quality assurance* Adoption of quality control and assurance procedures would reduce the uncertainty significantly.

(viii) *Training of staff* It is very important to train the field and laboratory staff in the measurement and recording protocols.

18.5 Quality Assurance, Quality Control and Verification

Quality control (QC), quality assurance (QA) and verification procedures are important components of the carbon inventory process, particularly in reducing the uncertainty involved in estimating carbon stocks and changes. IPCC (2003, 2006) provides the definitions and procedures for QA, QC and verification to enhance the transparency and accuracy of estimates that go into a carbon inventory.

- *Quality control* is a system of routine technical activities to measure and control the quality of the inventory as it is being developed, and is designed to
  - Provide routine and consistent checks to ensure data integrity, correctness and completeness of data.
  - Identify and address errors and omissions.
  - Document and archive inventory material and record all QC activities.
- *Quality assurance* is a planned system of review procedures conducted by personnel not directly involved in the inventory compilation/development process.
- *Verification* refers to a set of activities and procedures conducted during the planning and implementation of the carbon inventory methods and models, which would help to establish the reliability of the methods and procedures adopted. Verification refers to those procedures that are external to the inventory process and apply independent data and comparison with other similar studies and estimates.

QA/QC procedures need to be developed and adopted particularly for the following activities (Pearson et al. 2005b)

- Field measurements
- Sample preparation and laboratory measurements
- Data entry and analysis
- Data storage and management

18.5.1 Quality Control Procedures

General quality control procedures The general QC procedures include quality checks related to calculations, data processing, completeness and documenting and archiving procedures. Examples of QC activities and procedures include

- Checking the assumptions used in the models
- Checking the sampling procedure
- Checking for transcription errors in data inputs
- Checking the calculation procedures of carbon stocks and changes and units and conversions
- Checking the time trends in long-term monitoring for any outliers
- Reviewing internal documentation and archiving
- Checking the suitability of use of any default data or coefficients
- Checking the integrity of database files
  - Confirming that appropriate data processing steps are correctly represented in the database
  - Confirming that data relationships are correctly represented in the database

- ◦ Ensuring that data fields are properly labelled and have the correct design specifications
- ◦ Ensuring adequate documentation of database and model structure

18.5.2 Quality Assurance Review Procedures

Quality assurance comprises activities outside the actual carbon inventory processes. The estimates of carbon stocks and changes may be reviewed by external agencies in an unbiased way. It is important to involve experts or reviewers who were not involved in the carbon inventory estimations. The experts are required to assess the quality of the carbon inventory and to identify areas where improvements are necessary. Quality assurance procedure involves expert peer review covering

- Review of calculations and assumptions
- Review to ascertain whether the models used have undergone peer review
- Assessment of documentation of models, input data and other assumptions

18.5.3 Verification

Verification activities include comparing the estimates made in the project or activity under consideration with those reported in literature or made by other institutions for similar projects or activities. Verification activities would help the carbon inventory experts to identify the limitations and improve the accuracy of the estimates, especially if the estimates from the project or activity under consideration deviate widely from other estimates made for a similar project or activity. The verification procedure and approach would include comparing the following:

- Project or activity estimates of carbon stocks
- Methods and models adopted
- Sampling method and size
- Calculation procedures adopted
- Technical knowledge and capacity of the staff involved in field measurements and laboratory studies

18.6 Conclusions

Uncertainty of estimates of carbon stocks and changes in biological systems, such as forest land and grassland, is likely to be high due to large spatial variation in the carbon stocks, growth rates and losses. This spatial variation could be due to changes in species, soil, rainfall, topography and management practices. Therefore, it is important to understand the causes of uncertainty, estimate the uncertainty and adopt measures to reduce uncertainty. Identification of causes, and estimation and aggregation of errors in particular, require a detailed understanding of project features, key input parameters, models and the associated assumptions and statistical analysis techniques. The key factors to be considered in reducing uncertainty include

- Selection of appropriate sampling technique and size of the sample
- Selection of all the relevant parameters and adoption of correct measurement methods in the field and laboratory
- Selection of appropriate model for estimating carbon stocks and changes
- Adoption of quality assurance and quality control procedures

A good knowledge of the carbon inventory area or region, features of the vegetation, statistical tools and techniques, models and the associated assumptions and the human effort required for different activities of the inventory process is necessary. It is possible to use computer software, which can assist in detecting any errors, such as those in measurement and recording. An understanding of the sources and extent of uncertainty will contribute to reduction in human, material and financial resources required for reliable estimations. It is desirable to prepare a set of standard procedures and practices adopted in the project or national GHG inventory for quality control and assurance. Since any land use-based carbon inventory project or programme is a long-term activity, proper data storage and management are necessary, especially if the personnel are likely to change over the years. It is important to adopt quality control and assurance procedures described in this chapter and IPCC (2006) to ensure transparency and accuracy of the estimates.

Chapter 19
Implications of Climate Change for Carbon Stocks and Inventory

Climate change is one of the most important global environmental issues, which is likely to impact natural ecosystems as well as socio-economic systems (Ravindranath and Sathaye 2002). The global climate has warmed by 0.7 °C during the last century and is projected to rise by 1.8–4.0 °C during the current century (IPCC 2007a). This unprecedented projected rise in global mean temperature is likely to have adverse implications for forests, grassland and cropland vegetation. Climate is probably the most important determinant of vegetation patterns globally and has significant influence on the distribution, structure and ecology of forests (Kirschbaum et al. 1996). Climate vegetation studies have shown that certain climate regimes are associated with particular plant communities or forest types (Holdrige 1947). Thus, any change in the climate would alter the configuration of forest ecosystems (Kirschbaum 1996). Further, even a modest global warming of 1–2 °C will affect most ecosystems and landscapes through changes in species composition, productivity and biodiversity (Schröter et al. 2004). The coming decades will be characterized by increasing concentration of CO_2 in the atmosphere along with warming and changes in precipitation patterns. These factors may lead to changes in plant productivity. Such an impact on plant productivity will have implications for carbon inventory, since carbon stocks in forests, plantations and grasslands may increase or decrease in different regions. Implications for carbon inventory include the following:

- Changes in the size of the carbon sink in forest and other land-use systems
- Changes in growth rates of biomass and soil carbon in forests, grasslands and agroforestry systems
- Projections of carbon sequestration or mitigation potential and rates per hectare
- Greenhouse gas inventory or CO_2 emissions and removals in the long-term
- Estimation of timber or roundwood production

This chapter presents the projected impacts of climate change on carbon stocks and rates of change affecting the mitigation potential, particularly for forest and plantation systems, and implications for roundwood production, focusing on the implications for carbon inventory methods.

19.1 CO_2 Concentrations and Climate Change Projections

Projected CO_2 concentrations in the atmosphere The concentration of CO_2 in the atmosphere at the beginning of Industrial Revolution was about 280 ppm, which had increased to 380 ppm by 2005. The CO_2 concentration is projected to increase according to different scenarios in the range of 540–970 ppm by 2100 (Nakicenovic et al 2000). The main contributors to the increase are fossil-fuel combustion and deforestation, which increased from 1970 to 2004. The percent increase in emissions for different sectors included; 65% for industry, 120% for transport, 145% for energy supply while land use, land-use change and forestry increased with 40% (IPCC 2007c).

Recent evidence shows that CO_2 concentration increased by 1.9 ppm annually during 1995–2005, compared to the long-term average rate of 1.5 ppm since the beginning of the Industrial Revolution (IPCC 2007a). The growing concentration of CO_2 is likely to have positive implications for carbon sinks and net primary productivity so long as soil water or nutrients do not become the limiting factors.

Projected climate change Global average surface temperature during the last 100 years has increased by 0.74 °C, at the rate of 0.13 °C per decade over the last 50 years. The average global warming under various scenarios is estimated to be in the range of 1.8–4 °C, the best estimate being 2.4 °C by the end of the current century, with land surface warmer than the ocean surface. In addition to the warming, precipitation is likely to increase at higher latitudes and decrease in most subtropical land regions (IPCC 2007a). The combination of warming and changes in precipitation will impact the vegetation, particularly in terms of carbon stocks and plant productivity.

Nitrogen deposition Nitrogen deposition consists of the input of reactive nitrogen species from the atmosphere to the biosphere. A simulation study by Lamarque et al. (2005) showed that under the assumed IPCC SRES A2 scenario the global annual average nitrogen deposition over land is expected to increase by a factor of 2.5, mostly because of the increase in nitrogen emissions. On average, approximately 70% of the emitted nitrogen is deposited over the land masses. The results from this study suggest that the deposition over land ranges between 25 and 40 million tonnes (N) a year and, by 2100, will range between 60 and 100 million tonnes. Nitrogen deposition on forests is expected to double by 2100, which is likely to have implications for forest biomass stock and growth rates. The sources of nitrogen emissions include fossil fuel combustion, biomass burning, organic manure and application of nitrogenous fertilizer.

19.2 Impact of Climate Change on Forest Ecosystems

Forests account for nearly one third of the earth's land area with the tropics dominating by accounting for 42% of total global forest area. Forests store the largest portion of biospheric carbon stocks, estimated at 1,146 billion tonnes of carbon. A warming greater than 2–3 °C is projected to impact forest ecosystems (IPCC 2001, 2007b) in following ways.

- Significant forest dieback towards the end of the century in tropical, boreal and mountain areas, leading to loss in biodiversity, reduction in carbon sinks and loss of other forests services
- Substantial changes in structure and functioning of terrestrial ecosystems
- Nearly one third of the species assessed at increasing risk of extinction

Impact on forest biomass production and net primary productivity Scientific evidence of the impact of climate change on plant productivity is not yet conclusive. The projected climate change is likely to lead to gains and losses in plant productivity in different regions and over different periods.

- A review of impacts of climate change suggests varying levels of impacts (IPCC 2001, 2007b; Boisvenue and Running 2006). According to both field and satellite-based data, climate change over the last 55 years seems to have had a generally positive impact on forest productivity. However, this is only true of sites where water is not a limiting factor. Average productivity gains may result from CO_2 fertilization, although this effect is likely to be much weaker than previously projected, and plant productivity can also be limited by availability of nutrients.
- The atmospheric system has experienced changes not only in temperature but also in precipitation and solar radiation, in addition to rise in CO_2 concentration. Response of forest vegetation to rise in concentration of CO_2 is still uncertain, although studies (Wittig et al. 2005) found that gross primary productivity increased dramatically in the initial years at the younger state of development of species.
- A study by Ravindranath et al. (2007), using the equilibrium model BIOME, estimated that the net primary productivity under A2 and B2 IPCC-SRES scenarios would increase by 70–100% over the control or baseline scenario for different forest types in India, provided water or soil nitrogen are not the limiting factors.
- Carbon stock in the present-day vegetation is estimated to be about 600–630 GtC; between 2060 and 2100, it is predicted to increase by 290 GtC under the Had CM2 climate scenario and 170 GtC under the Had CM3 climate scenario (White et al 1999).
- Individual species responses to doubling of CO_2 can range from close to zero if nutrients, especially N, and water are the limiting factors (Oren et al. 2001; Reich et al. 2006) up to 70% when they are not (Morgan et al. 2004).
- Globally, forestry production is estimated to change only modestly with climate change in the short and medium term. The change in global forest products output could range from a modest increase to a slight decrease, although regional and local changes are likely to be large. Production increase is likely to shift from low-latitude regions in the short term to high-latitude regions in the long term.

There is little doubt that increasing atmospheric CO_2 can increase carbon uptake, and possibly carbon sinks, but the magnitude and spatial distribution of these impacts are still debated (Morgan et al. 2004; Körner et al. 2005). Climate Change coupled with increasing CO_2 concentration, nitrogen deposition and other factors, is likely to impact net primary productivity and biomass production in forest, plantation and grassland ecosystems. This will have implications for carbon inventory, in particular projection of future carbon stocks or biomass production in land-use systems.

19.3 Models for Projecting Carbon Stock Changes Under Climate Change Scenarios

Forests are highly complex biological ecosystems. Ecological modelling has been applied in forest science to make long-term projections of the impacts of climate change on forest vegetation dynamics. Predicting the effects of future climate change and human disturbances on the distribution of natural vegetation requires modelling. The models used to predict responses of vegetation to future climate change are categorized as follows and an illustrative list of models and their features is presented in Table 19.1.

Static biogeographical models Static biogeographical models assume equilibrium conditions in both climate and vegetation in predicting the distribution of potential vegetation by relating the geographic distribution of climatic parameters to vegetation. The equilibrium approach implicitly ignores the dynamic processes but generally requires far less information and estimates the potential magnitude of the response of vegetation from regional to global scales. These equilibrium models are restricted to the estimating steady-state conditions and include such models (Table 19.1) as BIOME and the mapped-atmosphere-plant–soil system model (MAPSS) (Woodward 1987).

Dynamic biogeographical models Dynamic biogeographical models capture the transient response of vegetation or biomes to a changing environment using explicit representation of key ecological processes such as establishment of a tree plantation, tree growth, competition, death and nutrient cycling (Shugart and West 1980; Shugart 1990; Botkin 1993). Dynamic models also require much more information on the characteristics of species than is easily available or even known for some areas of the globe (Solomon 1986). These models are used in predictions at the regional scale or for ecosystems, and have also been applied at the global scale. Examples of such models include HYBRID and IBIS (Table 19.1).

Process-based biogeochemistry models Biogeochemistry models project changes in basic ecosystem processes, such as the cycling of carbon, nutrients and water. These models are designed to predict changes in nutrient cycling and primary productivity. The inputs to these models are temperature, precipitation, solar radiation, soil texture and atmospheric CO_2 concentration. The plant and soil processes simulated are photosynthesis, decomposition, soil nitrogen transformations mediated by microorganisms, evaporation and transpiration. Common outputs from biogeochemistry models are estimates of net primary productivity, net nitrogen mineralization, evapotranspiration fluxes and the storage of carbon and nitrogen in vegetation and soil. Examples of such models are BIOME-BGC (Hunt and Running 1992; Running and Hunt 1993), CENTURY (CENTURY

Table 19.1 Features of selected climate impact assessment models and their relevance to carbon inventory

Model	Features	Data needs	Outputs	Application
Hybrid	Numerical process-based model; considers the daily cycling of C, N and H_2O within the biosphere and between the biosphere and atmosphere and projects C in soil and vegetation and productivity	- Number of wet days in each month - Mean precipitation per wet day - Monthly mean of maximum temperature recorded daily - Monthly mean of minimum temperature recorded daily - Solar irradiance, vapour pressure - Past and projected CO_2 concentrations - Species-level parameters (33) - Plot-level parameters (23) - Individual-level parameters (17) - Plant phenological parameters (2)	- Carbon in vegetation (kgC/m^2) - Carbon in soil (kgC/m^2) - Annual gross primary productivity (kgC/m^2/year) - Annual net primary productivity (kgC/m^2/year)	Projecting C stocks and rates of change
BIOME	Coupled carbon and water flux scheme, which determines the seasonal maximum LAI that maximizes NPP for any given PFT	- Monthly mean temperature, precipitation, and sunshine hours - Water-holding capacity of top 30 cm and 120 cm of soil. - Water conductivity indices	- Area changes in forest types - Changes in NPP	Projection of NPP
RothC	Terrestrial ecosystem biogeochemistry model	- Projected monthly air temperature, rainfall and evaporation. - Clay content, sampling depth, bulk density and inert carbon in soil	- Total C in soil	Projection of C in soil

C: carbon, N: nitrogen, LAI: leaf area index, NPP: net primary productivity, PFT: plant functional types

RothC: http://www.rothamsted.bbsrc.ac.uk/aen/carbon/rothc.htm

Hybrid: http://eco.wiz.uni-kassel.de/model_db/mdb/hybrid.html, BIOME:http://mercury.ornl.gov/metadata/ornldaac/html/rgd/daac.ornl.gov_data_bluangel_harvest_RGED_QC_metadata_models_biome4.html

1992), and RothC models (Coleman and Jenkinson 1995). BIOME-BGC predicts stocks and fluxes of carbon in both vegetation and soil whereas CENTURY and RothC predict stocks of carbon in soil.

19.4 Implications of Climate Change for Carbon Inventory

Climate change is important especially for long-term biomass production and carbon mitigation projects. Projections of future carbon stocks or biomass production are required during the project development phase. The implications of climate change for carbon inventory for different programmes and projects are given in Table 19.2. Karjalainen et al. (2003) have prepared a carbon inventory of European forest sector for the year 1990 and projected it for the year 2050 by incorporating the growth functions calibrated by process-based models into the European Forest Information Scenario Model (EFISCEN) framework. Total amount of carbon in

Table 19.2 Implications of climate change for carbon inventory

Carbon inventory programme or project	Estimates required	Implications for carbon inventory	Feasibility and reliability of projections of climate impacts
Carbon mitigation	- Biomass and soil carbon stock projections - Growth rates of biomass and soil carbon	- Estimating carbon gains or losses due to CO_2 fertilization and climate change - Significant implications for long-term afforestation and avoided deforestation projects - Need to exclude credits due to indirect effects of human action or global change	- Projection of impacts of climate change and CO_2 fertilization is feasible using dynamic global vegetation models - High uncertainty of projections at regional, species and project level
Biomass production	Roundwood (biomass) stocks and growth rates	Significant for long rotation timber projects for economic analysis	- Limitations exists with respect to assessment of combined effects of climate change, CO_2 concentrations and other factors
GHG inventory	Factoring out of CO_2 fertilization effects on carbon stock changes	Need for assignment of C credits and debits for projections of future carbon sink	Limitations of model prevent factoring out the effects of indirect human impacts

1990 was estimated to be 12.8Gt, with 94% of it in tree biomass and forest soil and 6% in wood products in use. Average total carbon stock is projected to be 35% higher after 60 years under the business-as-usual scenario. Average total carbon stocks are projected to be about 5% higher under a climate change scenario consisting of a mean temperature increase of 2.58 °C and annual precipitation increase of 5–15% between 1990 and 2050 than those under current climatic conditions.

19.5 Conclusions

Biomass production, carbon sequestration rates and carbon stocks in forest, plantation, grassland and agroforestry systems are influenced by climate parameters, carbon dioxide concentrations, nutrient supply, soil moisture status, management practices and other factors. This dependence makes projections of future biomass production or carbon sequestration rates too complex a process. Models help in making projections and are available for projecting the impacts of climate change and increase in CO_2 concentration on carbon stocks and net primary plant productivity of forest and plantation systems. These models require a whole range of parameters related to climate, plant physiology, soil, moisture and so on to be defined for each location and plant functional type. Most vegetation models make projections only for the plant functional types incorporated in the model. Thus, climate impacts cannot be assessed for all the diverse forest or plantation types occurring in a project location or even in a country. The currently available models have limitations in addressing the combined effects of climate, soil nutrients, water status and management practices. The projected climate changes, particularly those in temperature and precipitation and increase in concentration of CO_2 in the atmosphere will have impacts on biomass production and carbon sequestration. Such impacts are particularly relevant to long-term projects such as afforestation, reforestation and avoided deforestation. In future, impacts of climate change and elevated CO_2 concentration on land-use systems will become important for programmes and projects aimed at addressing climate change, particularly mitigation, as well as estimation of national greenhouse gas emissions. Future developments in methods and models may enable a better understanding of the implications of impacts of climate change, such as elevated CO_2 concentration and nitrogen deposition, on biomass production or carbon sequestration.

References

ALOS (Advanced Land Observation Satellite), 2006, *The ALOS Kyoto & Carbon Initiative – Science Plan v 2.0*. http://www.eorc.jaxa.jp/ALOS/kyoto/kyoto_index.htm accessed Feb 2007.

Amundson, R., Stern, L., and Raisden, T., 1998, The isotopic composition of soil and soil-respired CO_2, *Geodermatology*. 82:83–114.

Asia Forest Network, 2002, *Participatory rural appraisal for community forest management: Tools and techniques*, Asia Forest Network, Bohol, Philippines.

Aukland, L., Moura Costa, P., and Brown, S., 2003, A conceptual framework and its application for addressing leakage: The case of avoided deforestation, *Climate Policy*. 3:123–136.

Avery, T.E. and Burkhart, H.E., 1983, *Forest measurements*, McGraw-Hill series in forest resources, McGraw-Hill, New York.

Baldocchi, D., 2003, Assessing the eddy covariance technique for evaluating carbon dioxide exchange rates of ecosystems: Past, present and future, *Global Change Biology*. 9:479–492.

Baldocchi, D., Falge, E., Lianhong, G., Olson, R., Hollinger, D., Running, S., Anthoni, P., Bernhofer, C., Davis, K., Evans, R., Fuentes, J., Goldstein, A., Katul, G., Law, B., Xuhui, L., Malhi, Y., Meyers, T., Munger, W., Oechel, W., and Paw, U.K.T., 2001, FLUXNET: A new tool to study the temporal and spatial variability of ecosystem-scale carbon dioxide, water vapor and energy flux densities, *Bulletin of the American Meteorological Society*. 82:2415–2434.

Barret, E.C. and Curtis, L.F., 1995, *Introduction to environmental remote sensing*, Third edition, Chapman & Hall, London.

Baruah, T.C. and Barthakur, H.P, 1997, *A Textbook of Soil Analysis*, Vikas, New Delhi.

Bazilevich, N., 1974, *Geochemical work of the living matter of the Earth and soil forming*, in Pap. X International Congress of soil-scientists. Vol. 4, part 1, Nauka Press, Moscow, pp. 17–19 (in Russian).

Bock, M., Xofis, P., Mitchley, J., Rossner, G., and Wissen, M., 2005, Object-oriented methods for habitat mapping, Fat multiple scales – case studies from Northern Germany and Wye Downs, UK, *Journal for Nature Conservation*. 13:75–89.

Bohm, W., 1979, *Methods of studying roots systems*, Ecological studies No. 33, Springer, Berlin/Germany/New York.

Boisvenue, C. and Running, S.W., 2006, Impacts of climate change on natural forest productivity – evidence since the middle of the 20th century, *Global Change Biology*. 12:1–21.

Bolin, B. and Sukumar, R., 2000, Global perspective, in Watson, R.T., Noble, I.R., Bolin, B., Ravindranath, N.H., Verardo, D.J., and Dokken, D.J. (eds.), *IPCC. Land use, land-use change and forestry*, Special report, Cambridge University Press, Cambridge, Chap. 1.

Botkin, D.B., 1993, *Forest dynamics: an ecological model*, Oxford University Press, New York.

Brady, N.C. and Weil, R.R., 1999, *The Nature and properties of soils*, Prentice-Hall, Upper Saddle River, NJ.

Bickel, K., Richards, G., Köhl, M., and Rodrigues, R.L.V., 2006, Chapter 3 Consistent representation of land, in *IPCC Guidelines for National Greenhouse Gas Inventories*, Cambridge University Press, Cambridge.

Brogaard, S., Runnström, M., and Seaqusit, J.W., 2005, Primary production of Inner Mongolia, China, between 1982 and 1990 estimated by a satellite data-driven light use efficiency model, *Global and Planetary Change*. 45:313–332.

Brown, K., 1996, The utility of remote sensing technology in monitoring carbon sequestration agroforestry projects, Winrock International Working paper 7/96, Winrock International, Arlington, VA.

Brown, S., 1997, *Estimating biomass and biomass change of tropical forests: A primer*, FAO Forestry Paper 134, Food and Agriculture Organization of the United Nations, Rome, Italy.

Brown, S., Sathaye, J., Cannel, M., and Kauppi, P.E., 1996, Mitigation of Carbon Emission to the Atmosphere by Forest Management, *Commonwealth Forestry Review*. 75(1):80–91.

Brown, S.L. and Schroeder, P.E., 1999, Spatial patterns of aboveground production and mortality of woody biomass for eastern US forests, *Ecological Applications*. 9:968–980.

Cairns, M.A., Brown, S., Helmer, E.H., and Baumgardner, G.A., 1997, Root biomass allocation in the world's upland forests, *Oecologia*. 111:1–11.

Canadell, J.G., Mooney, H.A., Baldocchi, D.D., Berry, J.A., Ehleringer, J.R., Field, C.B., Gower, S.T., Hollinger, D.Y., Hunt, J.E., Jackson, R.B., Running, S.W. Shaver, G.R., Steffen, W., Trumbore, S.E., Valentini, R., and Bond, B.Y., 2000, Carbon metabolism of the terrestrial biosphere, *Ecosystems*. 3:115–130.

Canadell, J.G., Ciais, P., and Cox, P., 2004, Quantifying, understanding and managing the carbon cycle in the next decades, *Climatic Change*. 67:147–160.

Cannell, M.G.R., 1982, *World forest biomass and primary production data*, Academic Press, London.

Castillo-Santiago, M.A., Hellier, G., Tipper, R., and de Jong, B.H.J., 2006, Carbon emissions for land-use change: A regional analysis of causal factors in Chiapas, Mexico, *Mitigation and adaptation strategies for global change*. DOI 10.1007/s11027-006-9060-7.

CENTURY, 1992, *Soil organic matter model environment, technical documentation, agroecosystem version 4.0*, Colorado State University, Fort Collins, CO.

Chaturvedi, A.N. and Khanna, L.S., 1982, *Forest mensuration*, International Book Distributors, Dehradun.

Chaturvedi, R.K., 2007, *CES technical report*, Centre for Ecological Sciences, Indian Institute of Science, Bangalore.

Chomitz, K., 1998, *Baselines for greenhouse gas reductions: Problems, precedents, solutions*, Development Research Group, May, The World Bank, Washington, DC.

Clark, D.A., Brown, S., Kicklighter, D.W., Chambers, J.Q., Thomlinson, J.R., and Ni, J., 2001, Measuring net primary production in forests: Concepts and field measurements, *Ecological Applications*. 11:356–370.

Cole, V., Cerri, C., Mosier, A., Rosenberg, N.J., and Sauerbeck, D., 1996, Agricultural Options for Mitigation of Greenhouse Gas Emissions, Chapter 23, in Watson, R.T., Zimoywera, M.C., and Moss, R.H. (eds.), *Climate change: Impacts, adaptation and mitigation of climate change, Report of IPCC Working Group II*, Cambridge University Press, Cambridge.

Coleman, K. and Jenkinson, D.S., 1995, Roth C-26 3.A model for the turnover of carbon in soil: Model description and users guide. ISBN 0951-4456-69.

Commonwealth of Australia, 2001, *Bush for greenhouse, field measurement procedures for carbon accounting*, Australian Greenhouse office, Canberra.

Dale, V.H., Houghton, R.A., Grainger, A., Lugo, A.E., and Brown, S., 1993, Emissions of greenhouse gases from tropical deforestation and subsequent uses of the land, in Sustainable Agriculture and the Environment in the Humid Tropics, National Academy Press, Washington, DC.

Das, S. and Ravindranath, N.H., 2006, Remote sensing techniques for biomass production and carbon sequestration projects, in Rosillo-Calle, F. (ed.), *Biomass assessment handbook*, Earthscan, London.

Davidson, E.A. and Ackerman, I.L., 1993, Changes in soil carbon inventories following cultivation of previously untilled soils, *Biogeochemistry*. 20:161–193.

DeFries, R., Achard, F., Brown, S., Herold, M., Mudiyarso, D., Schlamadinger, B., and de Souza, Jr., C., 2006, *Reducing greenhouse gas emissions for deforestation in developing countries: Considerations for monitoring and measuring, GOFC-GOLD GTOS land cover global observation of forest and land cover dynamics*. GTOS 46, GOLD 26.

DeFries, R.G., Asner, F., Achard, C., Justice, N., Laporte, K., Price, C., and Townshend, J., 2005, Monitoring Tropical Deforestation for Emerging Carbon Markets, in Mountinho, P. and Schwartzman, S. (eds.), *Tropical deforestation and climate change*. IPAM, Belem, Brazil and Environmental Defense, Washington, DC.

De Vries, R.S., 1986, *Effects of land-use history on sedimentation in the Potomac Estuary, Maryland*. United States Geological Survey Water Supply Paper 2234-K, Washington, DC.

Delaney, M., Brown, S., and Powell, M., 1999, Carbon-Offset Report for the Noel-Kempff Project, Bolivia. *Report to nature conservancy*, Winrock International, Arlington, VA.

Dong, J., Kaufmann, R.T., Myneni, R.B., Tucker, C.J., Kauppi, P.E., Liski, J., Buermann, W., Alexeyev, V., and Hughes, M.K., 2003, Remote sensing estimates of boreal and temperate forest woody biomass: Carbon pools, sources, and sinks, *Remote Sensing of Environment*. 84:393–410.

Drake, J.B., Knox, R.G., Dubayah, R.O., Clark, D.B., Condit, R., Blair, J.B., and Hofton, M., 2003, Above-ground biomass estimation in closed canopy Neotropical forests using lidar remote sensing: Factors affecting the generality of relationships, *Global Ecology and Biogeography*. 12:147–159.

Eggleston, H.S., Salway, A.G., Charles, D., Jones, B.M.R., and Milne, R., 1998, *Treatment of uncertainties for national greenhouse gas emissions*, Report AEAT 2688–1 for DETR Global Atmosphere Division, AEA Technology, Culham.

FAO, 2004, *A review of carbon sequestration projects*. AGL/MISC/37/2004. Rome, Italy.

FAO, 2005, *Land cover classification system: Classification concepts and user manual*, Environment and Natural Resources Series, No. 8. Rome, Italy.

FAO, 2006, Global forest resources assessment 2005: Progress towards sustainable forest management, FAO Forestry Paper 147. FAO, Rome, Italy.

Fassnacht, K.S., Gower, S.T., MacKenzie, M.D., Nordheim, E.V., and Lillesand, T.M., 1997, Estimating the leaf area index of North Central Wisconsin forest using Landsat Thematic Mapper, *Remote Sensing of Environment*. 61:229–245.

Fishman, G.S., 1995, *Monte Carlo: Concepts, algorithms, and applications*, Springer Series in operation research, Springer, New York.

Forest Survey of India (FSI), 1996, *Volume equations for forests of India, Nepal and Bhutan*, Ministry of Environment and Forest, Govt. of India.

Franklin, J.F., 1989, Importance and justification of long-term studies in ecology, in Likens, G.E. (ed.), *Long-term studies in ecology*. Springer, New York.

Freese, F., 1962, *Elementary forest sampling*. USDA Agricultural Handbook No. 232, Washington, DC.

Fuentes, D.A., Gamon, J.A., Cheng, Y., Claudio, H.C., Qiu, H.-L., Mao, Z., Sims, D.A., Rahman, A.F., Oechel, W., and Luo, H., 2006, Mapping carbon and water vapour fluxes in a chaparral ecosystem using vegetation indices derived from AVIRIS, *Remote Sensing of Environment*. 103:312–323.

Geider, R.J., Delucia, E.H., Falkowski, P.G., Finzi, A.C., Grime, J.P., Grace, J., Kana, T.M., La Roche, J., Long, S.P., Osborne, B.A., Platt, T., Prentice, I.C., Raven, J.A., Schlesinger, W.H., Smetacek, V., Stuart, V., Sathyendranath, S., Thomas, R.B., Vogelmann, T.C., Williams, P., and Woodward, F.I., 2001, Primary productivity of planet earth: Biological determinants and physical constraints in terrestrial and aquatic habitats, *Global Change Biology*. 7:849–882.

Greenhouse, 2002, Field measurement procedures for carbon accounting, Bush for Greenhouse, Report No. 2. Version 1, Australian Greenhouse Office.

Hairiah, K., Sitompul, S.M., van Noordwijk, M., and Palm, C.A., 2001, Methods for sampling carbon stocks above and below ground, ASB Lecture Note 4B, International Centre for Research in Agroforestry (ICRAF), Bogor.http://www.icraf.cgiar.org/sea/Training/Materials/ASBTM/Datas/ASBLecNote%204B.pdf

Holdrige, L.R., 1947, Determination of world formation from simple climatic data, *Science*. 105:367–368.

Holmgren, J., Nilsson, M., and Olsson, H., 2003, Estimation of tree height and stem volume on plots using airborne laser scanning, *Forest Science*. 49:419–428.

Houghton, R.A., 1993, Changes in terrestrial carbon over the last 135 years, in Heimann, M. (ed.), *NATO ASI Series*, Vol. I 15: *The global carbon cycle*. Springer, New York, pp. 139–56.

Huete, A., Didan, K., Miura, T., Rodriguez, E.P., Gao, X., and Ferreira, L.G., 2002, Overview of radiometric and biophysical performance of MODIS vegetation indices, *Remote Sensing of Environment*. 83:195–213.

Hunt, E.R., Jr. and Running, S.W., 1992, Simulated dry matter yields for aspen and spruce stands in the North American boreal forest, *Canadian Journal of Remote Sensing*. 18:126–133.

Igarashi, T., Shimada, M., Rosenqvist, A., Hashimoto, T., Tadono, T., Matsuoka, M., and Yamamoto, H., 2003, Preliminary study on data sets of ADEOS-II and ALOS dedicated to terrestrial carbon observation, *Advances in Space Research*. 32:2147–2152.

IPCC (Intergovernmental Panel on Climate Change), 1996, Revised Guidelines. Cambridge University Press, Cambridge.

IPCC (Intergovernmental Panel on Climate Change), 2000a, *Good Practice Guidance and Uncertainty Management in National Greenhouse gas, Inventories*, Institute for Global Environmental Strategies, Hayama, Japan.

IPCC (Intergovernmental Panel on Climate Change), 2000b, *Land use, land-use change and forestry*, Special report, Cambridge University Press, Cambridge.

IPCC (Intergovernmental Panel on Climate Change), 2001a, *Climate change 2001: The scientific basis*, Summary for Policy Makers and Technical Summary of the Working Group I Report, Geneva, Switzerland.

IPCC (Intergovernmental Panel on Climate Change), 2001b, *Climate change 2001: The scientific basis*, Mitigation of Climate Change, Working Group III Report, Geneva, Switzerland.

IPCC (Intergovernmental Panel on Climate Change), 2003, *Good practice guidance for land use, land-use change and forestry*. Institute for Global Environmental Strategies, Hayama, Japan.

IPCC (Intergovernmental Panel on Climate Change), 2004, *Good practice guidance*. Cambridge University Press, Cambridge.

IPCC (Intergovernmental Panel on Climate Change), 2006, *Guidelines for national greenhouse gas inventories*. Vol. 4, Agriculture, forestry and other land use (AFLOLU). Institute for Global Environmental Strategies, Hayama, Japan.

IPCC (Intergovernmental Panel on Climate Change), 2007a, *Climate change 2007*: Working Group I, Fourth Assessment Report, Technical Summary, Geneva, Switzerland.

IPCC (Intergovernmental Panel on Climate Change), 2007b, *Climate change 2007*: Working Group II, Climate Change Impacts, Adaptation and Vulnerability, Technical Summary, Geneva, Switzerland.

IPCC (Intergovernmental Panel on Climate Change), 2007c, *Climate change 2007*: Working Group III, Mitigation of Climate Change, Technical Summary, Intergovernmental Panel on Climate Change, Geneva, Switzerland.

Jackson, R.B., Canadell, J., Ehleringer, J.R., Mooney, H.A., Sala, O.E., and Schulze, E.D., 1996, A global analysis of root distributions for terrestrial biomes, *Oecologia*. 108:389–411.

Joanneum Reserch, Union of Concerned Scientists, Woods Hole Research Centre and Instituto de Pesquisa Ambiental da Amazonia, 2006, *Reducing emissions from deforestation in Developing countries: Potential policy approaches and positive incentives*, Submission to the UNFCCC/SBSTA/2006/L.25.

Johnson, C.M., Zarin, D.J., and Johnson, A.H., 2000, Post-disturbance above-ground biomass accumulation in global secondary forests, *Ecology*. 81:1395–1401.

Kalara, Y.P. and Maynard, D.G., 1991, *Methods manual for forest soil and plant analysis*, Forestry Canada, Edmonton, Alberta.

Kale, M., Singh, S., Roy, P.S., Deosthali, V., and Ghole, V.S., 2004, Biomass equations of dominant species of dry deciduous forest in Shivpuri district, Madhya Pradesh, *Current Science*. 87:683–687.

Kangas, A. and Maltamo, M., 2006, *Forest inventory methodology and applications*, Vol. 10, Springer, Dordrecht, The Netherlands.

Karjalainen, T., Pussinen, A., Liski, J., Nabuurs, G.-J., Eggers, T., Lapvetelainen, T., and Kaipainen, T., 2003, Scenario analysis of the impacts of forest management and climate change on the European forest sector carbon budget, *Forest policy and Economics*. 5:141–155.

Kasischke, E.S., Melack, J.M., and Dobson, M.C., 1997, The use of imaging radar for ecological applications – a review, *Remote Sensing of Environment*. 59:141–156.

Kauppi, P., Sedjo, R., Apps, M., Cerri, C., Fujimori, T., Janzen, H., Krankina, O., Makundi, W., Marland, G., Masera, O., Nabuurs, G.J., Razali, W., and Ravindranath, N.H., 2001., Technical and economic potential of options to enhance, maintain and manage biological carbon reservoirs and geo-engineering, in *Climate change: Mitigation, Intergovernmental panel on climate change*, New York, Cambridge University Press, Cambridge.

Keeling, C.D. and Whorf, T.P., 1999, Atmospheric CO_2 records from sites in the SIO air sampling network, in *Trends: a compendium of data on global change*, Carbon Dioxide Information Analysis Center, Oak Ridge National Laboratory, Oak Ridge, TN.

Kimes, D.S., Ranson, K.J., Sun, G., and Blair, J.B., 2006, Predicting lidar measured forest vertical structure from multi-angle spectral data, *Remote Sensing of Environment*. 100:503–511.

Kirschbaum, M.U.F., Fischlin, A., Cannell, M.G.R., Cruz, R.V., Galinski, W., and Cramer, W.A., 1996, Climate Change Impacts on Forests, in Watson, R.T., Zinyowera, M.C., and Moss, R.H. (eds.), *Climate change 1995: Impacts, adaptations and mitigation of climate change: Second assessment report of the Intergovernmental Panel on Climate Change*, Cambridge University Press, Cambridge.

Körner, C., Asshoff, R., Bignucolo, O., Hättenschwiler, S., Keel, S.G., Pel'ez-Riedl, S., Pepin, S., Siegwolf, R.T.W., and Zotz, G., 2005, Carbon flux and growth in mature deciduous forest trees exposed to elevated CO2, *Science*. 309(5739):1360–1362.

Labrecque, S., Fournier, R., Luther, J., and Piercey, D., 2004, *A comparison for three approaches to map forest biomass from Landsat-TM and inventory data in Western Newfoundland*, Canadian Aeronautics and Space Institute Tasking ref 4–05.

Lal, R., Kimble, J.M., and Follett, R.F. (eds.), 2001, *Assessment methods for soil carbon*, CRC/Lewis Publishers, Boca Raton, FL.

Lamarque, J.-F., Kiehl, J.T., Brasseur, G.P., Butler, T., Cameron-Smith, P., Collins, W.D., Collins, W.J., Granier, C., Hauglustaine, D., Hess, P.G., Holland, E.A., Horowitz, L., Lawrence, M.G., McKenna, D., Merilees, P., Prather, M.J., Rasch, P.J., Rotman, D., Shindell, D., and Thornton, P., 2005, Assessing future nitrogen deposition and carbon cycle feedback using a multimodel approach: Analysis of nitrogen deposition, *Journal of Geophysical Research*. 110, D19303, DOI:10.1029/2005JD005825.

Lappi, J. and Kangas, A., 2006, Use of additional information, in Kangas, A. and Maltamo, M. (eds.), *Forest inventory, methodology and applications*. Springer, Dordrecht, The Netherlands.

Law, M.A. and Kelton, D., 2000, *Simulation modeling and analysis*, Third edition, McGraw-Hill, New York.

Lee, M.K., 2004, *CDM information and guidebook*. UNEP project CD4CDM, Roskilde, Denmark.

Lillesand, T.M. and Kiefer, R.W., 1994, *Remote sensing and image interpretation*. Wiley, New York.

Lillesand, T.M., Kiefer, R.W., and Chipman, J.W., 2004, *Remote sensing and image interpretation*, Fifth edition, Wiley, New York.

Lim, K.S. and Treitz, P.M., 2004, Estimation of above ground forest biomass from airborne discrete return laser scanner data using canopy-based quantile estimators, *Scandinavian Journal for Forest Research*. 19:558–570.

Liski, J., Palosuo, T., Peltoniemi, M., and Sievänen, R., 2005, Carbon and decomposition model Yasso for forest soils, *Ecological Modelling*. 189:168–182.

Loetsch, F., Zöhrer, F., and Haller, K.E., 1973, *Forest inventory*. Vol. 2. BLV Verlagsgesellschaft, Munich, Germany.

Lovell, J.L., Jupp, D.L.B., Culvenor, D.S., and Coops, N.C., 2003, Using airborne and ground-based ranging Lidar to measure canopy structure in Austrian forests, *Canadian Journal for Remote Sensing*. 29:607–622.

Lu, D., Mausel, P., Brondizio, E., and Moran, E., 2002, Above-ground biomass estimation of successional and mature forest using TM images in the Amazon Basin, Symposium on Geospatial Theory, Processing and Applications, Ottawa, Canada.

Lucas, R.M., Cronin, N., Lee, A., Moghaddam, M., Witte, C., and Tickle, P., 2006, Empirical relationship between AIRSAR backscatter and Lidar-derived forest biomass, Queensland, Australia, *Remote Sensing of Environment*. 100:407–425.

Lugo, A.E.J. and Brown, S., 1993, Management of tropical soils as sinks or sources of atmospheric carbon, *Plant and Soil*. 149:27–41.

MacDicken, K.G., 1997, *A guide to monitoring carbon storage in forestry and agroforestry projects*, Winrock International, Arlington, VA.

Macko, S.A., 1981, Stable nitrogen isotope rations as tracers of organic geochemical processes, Ph.D. thesis, University of Texas, Austin.

Majdi, H., 1996, Root sampling methods – applications and limitations of the minirhizotron technique, *Plant and Soil*. 185:255–258.

Mann, L.K., 1985, A regional comparison of carbon in cultivated and uncultivated alfisols and mollisols in the central United States, *Geoderma*. 36:241–253.

Schelhaas, M.J., van Esch, P.W., Groen, T.A., de Jong, B.H.J., Kanninen, M., Liski, J., Masera, O., Mohren, G.M.J., Nabuurs, G.J., Palosuo, T., Pedroni, L., Vallejo, A. and Vilén,, T., 2004, *CO2FIX V 3.1 – A modelling framework for quantifying carbon sequestration in forest ecosystems*. Wageningen, Alterra, Alterrarapport 1068.

MEA (Millennium Ecosystem Assessment), 2005, *Ecosystems and human well-being: Synthesis*. Island Press, Washington, DC (also available at www.millenniummassessment.org).

Mikkelsen, B., 1995, *Methods for development work and research – a guide for practitioners*, Sage, New Delhi.

Misra, R., 1972, A comparative study of net primary productivity of dry deciduous forest and grassland of Varanasi, India, in Golley, P.M. and Golley, F.B. (eds.), *Tropical ecology with and emphasis on organic production*, University of Georgia, Athens, Georgia.

Morgan, J.A., Pataki, D.E., Körner, C., Clark, H., Del Grosso, S.J., Grünzweig, J.M., Knapp, A.K., Mosier, A.R., Newton, P.C.D., Niklaus, P.A., Nippert, J.B., Nowak, R.S., Parton, W.J., Polley, H.W., and Shaw, M.R., 2004, Water relations in grassland and desert ecosystems exposed to elevated atmospheric CO_2, *Oecologia*. 140:11–25.

Moura-Costa, P.M., Stuart, M., Pinard, M., and Phillips, G., 2000, Elements of a certification system for forest-based carbon offset projects, *Mitigation and Adaptation Strategies for Global Change*. 5:39–50.

Mokany, K., Raison, R.J., and Prokushkin, A.S., 2006, Critical analysis of root-shoot ratios in terrestrial biomes, *Global Change Biology*. 12:84–96.

Murali, K.S. and Bhat, D.M., 2005, Biomass estimation equations for tropical deciduous and evergreen forests, *International Journal of Agricultural Resources, Governance and Ecology*. 4:81–92.

Murthy, I.K., Hegde, G.T., Sudha, P., and Ravindranath, N.H., 2006, Methodological issues in developing a community forestry mitigation project in Mancherial forest division of Andhra Pradesh, India, *Environmental Science and Policy*. 9:525–537.

Muukkonen, P. and Heiskanen, J., 2006, Biomass estimation over a large area based on stand wise forest inventory data and ASTER and MODIS satellite data: A possibility to verify carbon inventories, *Remote Sensing of Environment*. DOI: 10.1016/J.rse.2006.10.011.

Myeong, S., David, J.N., and Michael, J.D., 2006, A temporal analysis of urban forest carbon storage using remote sensing, *Remote Sensing of Environment*. 101:277–282.

Myers, W.L. and Shelton, R.L., 1980, *Survey methods for ecosystem management*. Wiley, New York.

Myneni, R.B., Tucker, C.J., Asrar, G., and Keeling, C.D., 1998, Inter-annual variations in satellite-sensed vegetation index data from 1981 to 1991, *Journal of Geophysical Research*. 103(D6):6145–6160.

Næsset, E., 2002, Predicting forest stand characteristics with airborne scanning laser using a practical two-stage procedure and field data, *Remote Sensing of Environment*. 80:88–99.

Nagler, P.L., Cleverly, J., Glenn, E., Lampkin, D., Huete, A., and Wan, Z., 2005, Predicting riparian evapotranspiration from MODIS vegetation indices and meteorological data, *Remote sensing of Environment*. 94:17–30.

Nakicenovic, N., Alcamo, J., Davis, G., de Vries, B., Fenhann, J., Gaffin, S., Gregory, K., Grübler, A., Yong Jung, T., Kram, T., Lebre La Rovere, E., Michaelis, L., Mori, S., Morita, T., Pepper, W., Pitcher, H., Price, L., Riahi, K., Roehrl, A., Rogner, H., Sankovski, A., Schlesinger, M., Shukla, P., Smith, S., Swart, R., van Rooijen, S., Victor, N., and Dadi, Z., 2000, in Nakicenovic, N. and Swart, R (eds.), *Emission scenarios: Special report of Working Group III of the 2000 Intergovernmental panel on climate change Network*, Cambridge University Press, Cambridge.

Noble, I., Apps, M., Houghton, R., Lashof, D., Makundi, W., Murdiyarso, D., Murray, B., Sombroek, W., and Valentini, R., 2000, in MacDicken, K. and Manning, M. (eds.) Chapter 2: *Implications of different definitions and generic issues, in IPCC 2000*, Land use, land-use change and forestry, Special Report, Cambridge University Press, Cambridge.

Ogle, S.M., Breidt, F.J., and Paustian, K., 2005, Agricultural Management Impacts on Soil Organic Carbon Storage under Moist and Dry Climatic Conditions of Temperate and Tropical Regions, *Biogeochemistry*. 72:87–121.

Ostwald, M., 2002, GIS-based support system for decision making of local forest control Illustrated examples from Orissa, India, *Environmental Management*. 30:35–45.

Ostwald, M. and Chen, D., 2006, Land-use change: Impact of climate variability and policies among small scale farmers in the Loess Plateau, China, *Land Use Policy*. 23:361–371.

Oren, R., Ellsworth, D.S., Johnsen, K.H., Phillips, N., Ewers, B.E., Maier, C., Schäfer, K.V.R., McCarthy, H., Hendrey, G., McNulty, S.G., and Katul, G.G., 2001, Soil fertility limits carbon sequestration by forest ecosystems in a CO_2-enriched atmosphere, *Nature*. 411:469–472.

Parresol, B.R., 1999, Assessing tree and stand biomass: A review with examples and critical comparisons, *Forest Science*. 45:573–593.

Patenaude, G., Milne, R., and Dawson, T.P., 2005, Synthesis of remote sensing approaches for forest carbon estimation: Reporting to the Kyoto Protocol, *Environmental Science and Policy*. 8:161–178.

Paustian, K., Andren, O., Janzen, H.H., Lal, R., Smith, P., Tian, G., Tiessen, H., Van Noodwijk, M., and Woomer, P.L., 1997, Agricultural Soils as a sink to mitigate carbon dioxide emissions, *Soil Use and Management*. 13:230–244.

Pearson, T.R.H., Brown, S., and Ravindranath, N.H., 2005a, Integrating carbon benefit estimates into GEF projects, UNDP/GEF Capacity development and adaptation group, Guidelines.

Pearson, T., Walker, S., and Brown, S., 2005b, *Sourcebook for land use, land-use change and forestry projects*, Winrock International, Arlington, VA.

Petit, J.R., Jouzel, J., Raynaud, D., Barkov, N.I., Barnola, J.-M., Basile, I., Bender, M., Chappellaz, J., Davis, M., Delaygue, G., Delmotte, M., Kotlyakov, V.M., Legrand, M., Lipenkov, V.Y., Lorius, C., Pépin, L., Ritz, C., Saltzman, E., and Stievenard, M., 1999, Climate and atmospheric history of the past 420,000 years from the Vostok ice core, Antarctica, *Nature*. 399:429–435.

Ponce-Hernandez, R., Koohafkan, P., and Antoine, J., 2004, Assessing carbon stocks and modelling win–win scenarios of carbon sequestration through land-use changes, FAO Call No. 631.4 (E), FAO, Rome, Italy.

Porté, A. and Bartelink, H.H., 2002, Modelling mixed forest growth: A review of models for forest management, *Ecological Modelling*. 150:141–188.

Ravindranath, N.H. and Hall, D.O., 1995, *Biomass, energy and environment – a developing country perspective from India*, Oxford University Press, Oxford, 1995.

Ravindranath, N.H. and Sathaye, J., 2002, *Climate change and developing countries*, Kluwer Academic, Dordrecht, The Netherlands.

Ravindranath, N.H., Murthy, I.K., Chaturvedi, R.K., Andrasko, K., and Sathaye, J.A., 2007, Carbon price-driven forestry mitigation in India by land classification, *Mitigation and Adaptation Strategies for Global Change*. DOI. 10.1007/s11027-006-9063-4.

Ramachandran, A.A., Jayakumar, S., Mohamed Haroon, A.R., and Bhaskaran, A., 2007, Carbon management in forest floor: An agenda of 21st century in Indian forestry scenario, *Indian Forester*. January 133:25–40.

Reich, P.B., Hobbie, S.E., Lee, T., Ellsworth, D.S., West, J.B., Tilman, D., Knops, J.M.H., Naeem, S., and Trost, J., 2006, Nitrogen limitation constrains sustainability of ecosystem response to CO2, *Nature*. 440:922–925.

Reyes, G., Brown, S., Chapman, J., and Lugo, A.E., 1992, *Wood densities of tropical tree species, USDA Forest Service*, General Technical Report SO-88, Southern Forest Experiment Station, New Orleans, LA.

Richter, D.D., Markenwitz, D., Trembore, S.E., and Wells, C.G., 1999, Rapid accumulation and turnover of soil carbon in aggrading forests, *Nature*. 400:56–58.

Rosenqvist, Å., Milne, A., Lucas, R., Imhoff, M. and Dobson, C., 2003, A review of remote sensing technology in support of the Kyoto Protocol. *Environmental Science and Policy*. 6:441–455.

Rosillo-Calle, F., de Groot, P., Hemstock, S.L., and Woods, J., 2006, *Biomass assessment handbook*. Earthscan, London.

Running, S.W. and Hunt, E.R.J., 1993, Generalization of a forest ecosystem process model for other biomes, Biome-BGC, and an application for global-scale models, in Ehleringer, J.R. and Field, C.B. (eds.), *Scaling physiological processes: Leaf to globe*. Academic Press, San Diego, CA.

Running, S.W., Baldocchi, D.D., Turner, D.P., Gower, S.T., Bakwin, S., and Hibbard, K.A., 1999, A global terrestrial monitoring network, scaling tower fluxes with ecosystem modeling and EOS satellite data, *Remote Sensing Environment*. 70:108–127.

Rypdal, K. and Winiwarter, W., 2001, Uncertainties in Greenhouse Gas Inventories: Evaluation, comparability and implications, *Environmental Science and Policy*. 4:107–116.

Santantonio, D., Hermann, R.K., and Overton, W.S., 1997, Root biomass studies in forest ecosystems, *Pedobiologia*. 17:1–31.

Sanz, D.C., Caneill, J.-Y., de Lannoy, R., Markussen, P., and Rivero, C., Scowcroft, J., De Janeiro, V., 2005, *Manual on Joint Implementation (JI) and Clean Development Mechanism (CDM) projects*, Union of the electricity industry, Ref: 2005-030-303, Brussels.

Sathaye, J. and Mayers, S., 1995, *Greenhouse gas mitigation assessment: A guidebook*, Kluwer, Dordrecht, The Netherlands.

Sathaye, J. and Ravindranath, N.H., 1998, Climate Change Mitigation in the Energy and Forest Sectors of Developing Countries, *Annual Review of Energy and Environment*. 23:387–437.

Sathaye, J.A. and Andrasko, K., 2006, Special issue on estimation of baseline and leakage in carbon mitigation forestry projects, *Mitigation and Adaptation Strategies for Global Change*. DOI 10.1007/s11027-006-9057-2.

Schelhass, M.J., van Esch, P.W., Groen, T.A., de Jong, B.H.J., Kanninen, M., Liski, J., Masera, O., Mohren, G.M.J., Nabuurs, G.J., Palosuo, T., Pedroni, L., Vallejo, A., Vilen, T., 2004. CO2FIX V3.1.description of a model for quantifying carbon sequestration in forest ecosystems and wood products. ALTERRA Report 1068. ALTERRA, Wageningen.

Schmid, H.P., 1994, Source areas for scalars and scalar fluxes, *Boundary Layer Meteorology*. 67:293–318.

Schröter, D., Metzger, M.J., Cramer, W., and Leemans, R., 2004, Vulnerability assessment: Analysing the human–environment system in the face of global environmental change, *Environmental Science Section Bulletin*. 2:11–17.

SCOPE, 2006, The Global Carbon Cycle, *Policy briefs*, UNESCO, No. 2.

Seaquist, J.W., Olsson, L., and Ardö, J., 2003, A remote sensing-based primary production model for grassland biomes, *Ecological Modelling*. 169:131–155.

Shiver, B.D. and Borders, B.E., 1996, *Sampling techniques for forest resource inventory*. Wiley, New York.

Shugart, H.H., 1990, Using ecosystem models to assess the potential consequences of global climatic change, *Trends in Ecological Evolution*. 5:303–307.

Shugart, H.H. and West, D.C., 1980, Forest succession models, *Bioscience*. 30:308–313.

Skutsch, M., Bird, N., Trines, E., Dutschke, M., Frumhoff, P., de Jong, B., van Laake, P., Masera, O., and Mudiyarso, D., 2007, Clearing the way for reducing emissions from tropical deforestation, *Environmental Science and Policy*. 20:322–334.

Solomon, A.M., 1986, Transient response of forest to CO_2-induced climate change: Simulation modeling experiments in eastern North America, *Oecologia*. 68:567–579.

Steininger, M.K., 2000, Satellite estimation of tropical secondary forest above-ground biomass: Data from Brazil and Bolivia, *International Journal of Remote Sensing*. 21:1139–1157.

Tamás, J. and Lénárt, C., 2006, Analysis of a small agricultural watershed using remote sensing techniques, *International Journal of Remote Sensing*. 27:3727–3738.

Tan, K., Piao, S., Peng, C., and Fang, J., 2007, Satellite-based estimation of biomass carbon stocks for northeast China's forests between 1982 and 1999, *Forest Ecology and Management*. DOI. 10.1016/j.foreco.2006.12.018.

Todd, S.W., Hoffer, R.M., and Milchunas, D.G., 1998. Biomass estimation on grazed and ungrazed rangelands using spectral indices, *International Journal of Remote Sensing*. 19:427–438.

Tomppo, E., 2006, The Finnish multi-source national forest inventory: Small area estimation and map production, in Kangas, A. and Maltamo, M. (eds.), *Forest Inventory, Methodology and applications*, Springer, Dordrecht, The Netherlands.

Ugalde, X. and Perez, X., 2001, *Mean annual volume increment of selected industrial forest plantation species*, FAO, Rome, Italy.

UNEP (United Nations Environment Programme), 2002, *Global Environment outlook 3: Past, present and future perspectives*, Earthscan, London/Sterling, VA.

UNFCCC, 2002, *Methodological issues land use, land-use change and forestry: Definitions and modalities for including afforestation and reforestation activities under article 12 of the Kyoto Protocol in the first commitment period: Options paper on modalities for addressing baselines, additionality and leakage*, Bonn.

UNFCCC, 2006, *Background paper for the workshop on reducing emissions from deforestation in developing countries*, working paper No. 1.

Usher, M.B., 1991, Scientific requirements of monitoring programme, in Goldsmith, F.B. (ed.), *Monitoring for conservation*, Chapman & Hall, London.

Vine, E. and Sathaye, J., 1999, *Guidelines for the Monitoring, Evaluation, Reporting, Verification and Certification of Energy-Efficiency Projects for Climate Change Mitigation*. LBNL-41543, Lawrence Berkeley National Laboratory, Berkeley, CA.

Watson, R., Noble, I.R., Bolin, B., Ravindranath, N.H., Verardo, D.J., and Dokken, D.J., 2000, *Land use, land-use change, and forestry, IPCC special report*, Cambridge University Press, Cambridge.

WBGU, World in transition: Strategies for managing Global Environmental Risks German Advisory Council on Global Change *Annual Report 1998*, 2000, Springer, Berlin.

Weaver, J.E. and Darland, R.W., 1949, Quantitative study of root systems in different soil types, *Science*. 110:164–165.

Wenger, K.F. (ed.), 1984, *Forestry handbook*, Second edition, Wiley, New York.

White, A., Cannel, M.G.R., and Friend, A.D., 1999, Climate change impacts on ecosystems and the terrestrial carbon sink: A new assessment, *Global Environment Change*. 9:S3-S19.

Whittaker, R.H. and Likens, G.E., 1973, Carbon in the biota, in Woodwell, G.M. and Pecan, E.V. (eds.), *Carbon and the biosphere*, CONF 720510. US National Technical Information Service, Springfield, VA, pp. 281–302.

Winiwarter, W. and Rypdal, K., 2001, Assessing the uncertainty associated with national greenhouse gas emission inventories: A case study for Austria, *Atmospheric Environment*. 35:5425–5440.

Wittig, V.E., Bernacchi, C.J., Zhu, X.-G., Calfapietra, C., Ceulemans, R., Deangelis, P., Gielen, B., Miglietta, F., Morgan, P.B., and Long, S.P., 2005, Gross primary production is stimulated for three Populus species grown under free-air CO_2 enrichment from planting through canopy closure, *Global Change Biology*. 11:644–656.

Wofsy, S.C., Goulden, M.L., and Munger, J.W., 1993, Net exchange of CO2 in a mid-latitude forest, *Science*. 260:1314–1317.

Woodward, F.I., 1987, *Climate and plant distribution*. Cambridge University Press, Cambridge.

Wylie, B.K., Fosnight, E.A., Gilmanov, T.G., Frank, A.B., Morgan, J.A., Haferkamp, M.R., and Meyers, T.P., 2007, Adaptive data-driven model for estimating carbon fluxes in the Northern Great Plains, *Remote Sensing of Environment*. 106:399–413.

Zheng, D., Rademacher, J., Chen, J., Crow, T., Bresee, M., Le Moine, J. and Ryu, S.-R., 2004, Estimating aboveground biomass using Landsat 7 ETM + data across a managed landscape in northern Wisconsin, USA, *Remote Sensing of Environment*. 93:402–411.

List of web sites

www.CIFOR.org
http://www.ipcc-nggip.iges.or.jp/EFDB/main.php
www.unfccc.int

www.globalcarbonproject.org/budget.htm
www.undp.org
www.cdm.unfccc.int
www.fao.org
www.efdb.int
www.navcen.uscg.gov
www.winrock.org
http://www.rothamsted.bbsrc.ac.uk/aen/carbon/rothc.htm
http://eco.wiz.uni-kassel.de/model_db/mdb/hybrid.html
http://mercury.ornl.gov/metadata/ornldaac/html/rgd/daac.ornl.gov_data_bluangel_harvest_RGED_QC_metadata_models_biome4.html
http://earth.esa.int/dataproducts/
http://edc.usgs.gov
http://glcf.umiacs.umd.edu/data/gimms/
http://www. Eorc.nasda.go.jp/JERS-1
http://www.digitalglobe.com
http://www.esa.int
http://www.rsi.ca
http://www.spaceimaging.com
http://www.spot.com
http://www.spotimage.fr/home

Index

Advances in Global Change Research

1. P. Martens and J. Rotmans (eds.): *Climate Change: An Integrated Perspective.* 1999 ISBN 0-7923-5996-8
2. A. Gillespie and W.C.G. Burns (eds.): *Climate Change in the South Pacific: Impacts and Responses in Australia, New Zealand, and Small Island States.* 2000 ISBN 0-7923-6077-X
3. J.L. Innes, M. Beniston and M.M. Verstraete (eds.): *Biomass Burning and Its Inter-Relationships with the Climate Systems.* 2000 ISBN 0-7923-6107-5
4. M.M. Verstraete, M. Menenti and J. Peltoniemi (eds.): *Observing Land from Space: Science, Customers and Technology.* 2000 ISBN 0-7923-6503-8
5. T. Skodvin: *Structure and Agent in the Scientific Diplomacy of Climate Change.* An Empirical Case Study of Science-Policy Interaction in the Intergovernmental Panel on Climate Change. 2000 ISBN 0-7923-6637-9
6. S. McLaren and D. Kniveton: *Linking Climate Change to Land Surface Change.* 2000 ISBN 0-7923-6638-7
7. M. Beniston and M.M. Verstraete (eds.): *Remote Sensing and Climate Modeling: Synergies and Limitations.* 2001 ISBN 0-7923-6801-0
8. E. Jochem, J. Sathaye and D. Bouille (eds.): *Society, Behaviour, and Climate Change Mitigation.* 2000 ISBN 0-7923-6802-9
9. G. Visconti, M. Beniston, E.D. Iannorelli and D. Barba (eds.): *Global Change and Protected Areas.* 2001 ISBN 0-7923-6818-1
10. M. Beniston (ed.): Climatic Change: *Implications for the Hydrological Cycle and for Water Management.* 2002 ISBN 1-4020-0444-3
11. N.H. Ravindranath and J.A. Sathaye: *Climatic Change and Developing Countries.* 2002 ISBN 1-4020-0104-5; Pb 1-4020-0771-X
12. E.O. Odada and D.O. Olaga: *The East African Great Lakes: Limnology, Palaeolimnologyand Biodiversity.* 2002 ISBN 1-4020-0772-8
13. F.S. Marzano and G. Visconti: *Remote Sensing of Atmosphere and Ocean from Space: Models, Instruments and Techniques.* 2002 ISBN 1-4020-0943-7
14. F.-K. Holtmeier: *Mountain Timberlines.* Ecology, Patchiness, and Dynamics. 2003 ISBN 1-4020-1356-6
15. H.F. Diaz (ed.): *Climate Variability and Change in High Elevation Regions: Past, Present & Future.* 2003 ISBN 1-4020-1386-8
16. H.F. Diaz and B.J. Morehouse (eds.): *Climate and Water: Transboundary Challenges in the Americas.* 2003 ISBN 1-4020-1529-1
17. A.V. Parisi, J. Sabburg and M.G. Kimlin: *Scattered and Filtered Solar UV Measurements.* 2004 ISBN 1-4020-1819-3
18. C. Granier, P. Artaxo and C.E. Reeves (eds.): *Emissions of Atmospheric Trace Compounds.* 2004 ISBN 1-4020-2166-6
19. M. Beniston: *Climatic Change and its Impacts.* An Overview Focusing on Switzerland. 2004 ISBN 1-4020-2345-6
20. J.D. Unruh, M.S. Krol and N. Kliot (eds.): *Environmental Change and its Implications for Population Migration.* 2004 ISBN 1-4020-2868-7
21. H.F. Diaz and R.S. Bradley (eds.): *The Hadley Circulation: Present, Past and Future.* 2004 ISBN 1-4020-2943-8

Advances in Global Change Research

22. A. Haurie and L. Viguier (eds.): *The Coupling of Climate and Economic Dynamics.* Essays on Integrated Assessment. 2005 ISBN 1-4020-3424-5
23. U.M. Huber, H.K.M. Bugmann and M.A. Reasoner (eds.): *Global Change and Mountain Regions.* An Overview of Current Knowledge. 2005 ISBN 1-4020-3506-3; Pb 1-4020-3507-1
24. A.A. Chukhlantsev: *Microwave Radiometry of Vegetation Canopies.* 2006 ISBN 1-4020-4681-2
25. J. McBeath, J. Rosenbery : *Comparative Environmental Politics.* 2006 ISBN 1-4020-4762-2
26. M.E. Ibarrarán and R. Boyd: *Hacia el Futuro.* Energy, Economics, and the Environment in 21st Century Mexico. 2006 ISBN 1-4020-4770-3
27. N.J. Roserberg: *A Biomass Future for the North American Great Plains:* Toward Sustain-able Land Use and Mitigation of Greenhouse Warming. 2006 ISBN 1-4020-5600-1
28. V. Levizzani, P. Bauer and F.J. Turk (eds.): *Measuring Precipitation from Space.* EURAINSAT and the Future. 2007 ISBN 978-1-4020-5834-9
29. N.H. Ravindranath and M. Ostwald: *Carbon Inventory Methods.* Handbook for Greenhouse Gas Inventory, Carbon Mitigation and Roundwood Production Projects. 2008 ISBN 978-1-4020-6546-0